# EMISSIONS TRADING
## *Principles and Practice*

## Second Edition

### T. H. Tietenberg

RESOURCES FOR THE FUTURE
WASHINGTON, DC, USA

An RFF Press book
Published by Resources for the Future
1616 P Street NW
Washington, DC 20036–1400
USA
www.rffpress.org

*Library of Congress Cataloging-in-Publication Data*

Tietenberg, Thomas H.
  Emissions trading : principles and practice / T. H. Tietenberg.-- 2nd ed.
    p. cm.
  Rev. ed. of: Emissions trading, an exercise in reforming pollution policy. 1985.
  Includes bibliographical references and index.
  ISBN 1-933115-30-0 (cloth : alk. paper) -- ISBN 1-933115-31-9 (pbk. : alk. paper)
  1. Emissions trading--United States. 2. Air--Pollution--Government policy--United States. 3. Environmental impact charges--United States. I. Tietenberg, Thomas H. Emissions trading, an exercise in reforming pollution policy. II. Title.

  HC110.A4T54 2006
  363.739'260973--dc22                                          2005033482

The paper in this book meets the guidelines for permanence and durability of the Committee on Production Guidelines for Book Longevity of the Council on Library Resources. This book was typeset by Peter Lindeman. It was copyedited by Patricia Miller. The cover was designed by Circle Graphics.

ISBN 1-933115-30-0  (cloth)          ISBN 1-933115-31-9  (paper)

# Contents

# Figures and Tables

# Dedication

*This book is dedicated to
the memory of
Allen V. Kneese
1930–2001*

*Pathbreaking scholar and one of the principal founders
of the field of environmental economics.*

# Preface

This second edition of *Emissions Trading* represents the culmination of more than 30 years of study about the role of tradable permit systems in environmental policy. My interest in this topic first arose during my graduate school years in the late 1960s and early 1970s. Inspired by the work of Allen V. Kneese and his colleagues at Resources for the Future (RFF), I became an environmental economist and in my dissertation crafted a general equilibrium model to evaluate the properties of various economic incentive approaches to pollution control, including tradable permits.

Despite the fact that an accumulating body of research at RFF and elsewhere seemed to suggest that building economic incentives into environmental policy could well offer promising opportunities for achieving better environmental and economic outcomes, during that time policy interest in any economic incentive system, including emissions trading, was nil. As the laws of physics tell us, a body at rest tends to stay at rest in the absence of a sufficiently powerful outside force. The fact that the advice was coming from academics (not practitioners) and economists (not lawyers) may have diminished, or at least delayed, its influence.

This disconnect between the academic and policy communities did not last long, however, as in the mid-1970s the U.S. Environmental Protection Agency began implementing some aspects of what subsequently became known as the Emissions Trading Program. In retrospect, on the basis of its outcomes, I think most of us would now judge that program as, at best, a very limited success and, at worst, a failure. In another sense, however, it was an important milestone, because, as the cliché goes, the camel's nose was now under the tent. Even failures have lessons to teach and, if they are not large enough to kill further interest, they can provide the foundation for what follows.

In this case, of course, that is exactly what happened. In 1984–1985, I was extremely fortunate to have been awarded a Gilbert White Fellowship at RFF, which allowed me to spend a year analyzing and writing about what was then a very new venture in environmental policy. The first edition of this book, *Emissions Trading: An Exercise in Reforming Pollution Policy*, was the result of that intellectually stimulating year.

Much has happened since that book was published in 1985. The tradable permits approach now has been extended to new pollutants including, notably, greenhouse gases, and to new geographic areas including, among others, the European Union and Santiago, Chile, and even to new, related areas of environmental protection, including the promotion of renewable energy sources. All of this developing implementation experience as well as the theoretical sophistication contributed by a whole new generation of scholars made the urge to bring that book up to date irresistible.

A full-year sabbatical from Colby College in 2004–2005 offered just the opportunity I needed to absorb and make sense of this exploding field, and this book is the result. In it, I have tried to capture the central lessons about emissions trading that have emerged to date from the theoretical and empirical research, as well as the implementation experience.

The picture that emerges, not surprisingly, is positive but mixed. Emissions trading certainly has accumulated some clear, impressive successes and because of those successes probably has irreversibly carved out a niche for itself in modern pollution control policy. The story also, however, uncovers many weaknesses of emissions trading, particularly in specific contexts, and this approach still faces many challenges that are as yet unresolved. I have tried to be equally clear about both its successes and the remaining challenges.

Readers of this book also may be interested in a rather extensive bibliography on tradable permits that is available on my web site at: http://www.colby.edu/~thtieten/trade.html. This bibliography contains sources dealing with subjects as diverse as the historical and theoretical foundations of emissions trading, design considerations, ethical dimensions, and studies that examine specific applications in air, water pollution control, energy, forestry, and fisheries, among others.

This book was improved significantly by the extremely helpful comments on the first draft that I received from Roberton C. Williams III, Thomas Sterner, Denny Ellerman, and Juan-Pablo Montero. I am deeply grateful for their assistance.

Tom Tietenberg
Colby College
Waterville, Maine

# Abbreviations

| | |
|---|---|
| ATS | Allowance Tracking System |
| CEM | Continuous emissions monitoring |
| CERs | Certified emission reductions |
| EPA | U.S. Environmental Protection Agency |
| ERMS | Emissions Reduction Market System |
| ETP | Emissions Trading Program |
| EU ETS | European Union Emissions Trading Scheme |
| NBP | $NO_x$ Budget Trading Program |
| OTC | Ozone Transport Commission |
| RACT | Reasonably available control technology |
| RECLAIM | Regional Clean Air Incentives Market |
| RTC | RECLAIM tradable credits |
| SIP | State implementation plan |
| UNFCCC | United Nations Framework Convention on Climate Change |
| VOC | Volatile organic chemical |
| VOM | Volatile organic matter |

# 1

# Introduction

A long tradition in economics suggests that treating resources as a commons that are shared jointly by many users could lead to overexploitation in the absence of some kind of access rationing (Ostrom et al. 2002). Since the atmosphere is one such commons, it is not surprising to find that in the absence of some kind of access rationing for polluters, the atmosphere would be excessively polluted. The policy question, therefore, is what form should the control over access take?

One increasingly common form involves the use of emissions trading. In contrast to more traditional regulation, where the regulatory authority specifies a specific maximum level of emissions for each emissions source within a plant, emissions trading is a regulatory program that allows pollution emitters considerable flexibility in how they comply with the regulation. With emissions trading, as long as the total emissions reduction is the same or greater, firms can comply by either: (1) reducing emissions from any combination of sources within the plant; or (2) acquiring emissions reductions from another facility.

The logic behind the growing prominence of this approach is simple. One of the insights derived from the empirical literature is that traditional command-and-control regulatory measures, which depend upon government agencies to define not only the goals but also the means for reaching them, are in many cases insufficiently protective of those resources or economically inefficient. Emissions trading provides, at least in principle, a cost-effective alternative.

Applications of this general approach have spread not only to many different types of pollution in many different countries but are also being used to ration access to many other resources, including fisheries, forests, water, and land use control, among others.[1] Though the lessons from emissions trading are certainly useful for those considering this approach for other resources, this book will focus on the use of this technique to control air pollution.

## Early History

By the late 1950s, both economists and policymakers had formed well-developed and deeply entrenched visions of how pollution control policy should be conducted. Unfortunately, the two visions were worlds apart.

Economists viewed the world through the eyes of Pigou (1920). Professor A.C. Pigou had argued that in the face of an externality such as pollution, the appropriate remedy involved imposing a per-unit tax on the emissions from a polluting activity. The tax rate would be equal to the marginal external social damage caused by the last unit of pollution at the efficient allocation. Faced with this tax on emissions, firms would "internalize" the externality. By minimizing their private costs, firms would simultaneously minimize the costs to society as a whole. According to this view, rational pollution control policy involved putting a price on pollution.

Policymakers, particularly, but not exclusively, in the United States, held an equally firm, if substantially different, view. According to this view, the proper way to control pollution was through a series of legal regulations ranging from controlling the location of polluting activities (to keep them away from people) to the specification of emissions ceilings (to limit the amount ejected into the air). Under this regulatory regime, the public sector would be responsible for: (1) figuring out how much pollution to allow each emitter (usually by identifying the specific control technology that should be used); (2) mandating either a specific technology or level of emissions flow achievable by that technology; (3) monitoring emissions to verify compliance with these mandates; and (4) using financial penalties or other sanctions to bring non-complying sources into compliance.

While some exchanges of ideas took place between the two groups, most of it was highly critical and not viewed by the recipients as particularly helpful. Economists would point out, for example, that legal regimes, which became known as "command-and-control" regimes, generally were not cost-effective. Hence, they argued that by simply switching to Pigouvian taxes, more pollution control could be gained with the same expenditure or the same pollution control could be gained with less expenditure.

Policymakers responded that the information burden imposed on the bureaucracy by the design of efficient taxes was unrealistically high. And taxes based upon very limited information might not be any better than legal regulations. Furthermore, they argued, if bureaucrats had sufficient information to set efficient tax rates, they could use the same information to set efficient legal regimes.

The result was a standoff in which policymakers focused on quantity-based policies while economists continued to promote price-based remedies. While the standoff continued, legal regimes prevailed. Taxes made little headway, particularly in the United States.

In 1960, Ronald Coase published a remarkable article in which he sowed the seeds for a different mind-set. Arguing that Pigou had used an excessively narrow focus, Coase went on to suggest:

> It is my belief that the failure of economists to reach correct conclusions about the treatment of harmful effects cannot be ascribed simply to a few slips in analysis. It stems from basic defects in the current approach to problems of welfare economics. What is needed is a change in approach. (1960, 42)

His proposed change in approach involved refocusing on property rights:

> If factors of production are thought of as rights, it becomes easier to understand that the right to do something which has a harmful effect (such as the creation of smoke, noise, smell, etc.) is also a factor of production.... The cost of exercising a right (of using a factor of production) is always the loss that is suffered elsewhere in consequence of the exercise of that right—the inability to cross land, to park a car, to build a house, to enjoy a view, to have peace and quiet or to breathe clean air. (1960, 44)

Coase argued that by making these property rights explicit and transferable, the market could play a substantial role. To his fellow economists, Coase pointed out that a property rights approach allowed the market to value the property rights, as opposed to the government in the Pigouvian approach. To policymakers, Coase pointed out that control regimes based purely on emissions limits provided no means for the rights to flow to their highest valued use.

It remained for this key insight to become imbedded in practical programs for controlling pollution. Dales (1968) pointed out its applicability for water and Crocker (1966) for air. Among his other contributions, Dales noted that the legal regimes imposed by the government for pollution control in fact had already established a property right in the right to emit. Unlike the property right system envisioned by Coase, however, this property right was not efficient because it was not transferable:

> The "regulatory" branches of modern governments create an enormous variety of valuable property rights that are imperfectly transferable, and that tend to be capitalized and monetized in ways that are usually unsuspected by their creators. (1968, 796)

One possibility to correct that inefficiency, of course, would be to make the existing system of property rights transferable. In a section that foreshadows much of what was to come, Dales (1968, *801*) suggested a means for doing this:

> The government's decision is, let us say, that for the next five years no more than x equivalent tons of waste per year are to be discharged into the waters of region A. Let it therefore issue x pollution rights and put them up for sale, simultaneously passing a law that everyone who discharges one equivalent ton of waste into the natural water system during a year must hold one pollution right throughout the year. Since x is less than the number of equivalent tons of waste being discharged at present, the rights will command a positive price—a price sufficient to result in a 10 percent reduction in waste discharge. The market in rights would be continuous. Firms that found that their actual production was likely to be less than their initial estimate of production would have rights to sell, and those in the contrary situation would be in the market as buyers. Anyone should be able to buy rights; clean-water groups would be able to buy rights and not exercise them. A forward market in rights might be established.... The virtues of the market mechanism are that no person, or agency, has to *set* the price—it is set by the competition among buyers and sellers of rights.

In his discussion of how this approach could be used to control air pollution, Crocker (1966, *81*) noted a more basic point, namely that this approach fundamentally changes the information requirements imposed on the bureaucracy:

> Although the atmospheric pollution control authority's responsibilities will continue to be a good deal broader than the basic governmental function of providing legal and tenure certainty in property rights, its necessary work will not have to include the guesswork involved in attempting to estimate individual emitter and receptor preference functions.

When emissions trading is used to pursue a predetermined goal that specifies the level of allowable emissions, the authority does not have to know anything about either damage or cost functions.[2] Transferability, at least in principle, allows the market to handle the task of ensuring that the assignment of control responsibility ultimately ends up being placed on those who can accomplish the previously stipulated reductions at the lowest cost.

The final stage in this evolution was reached with the publication of a couple of now classic articles. The first, by Baumol and Oates (1971), formalized the theory behind these practical insights for the case of uniformly mixed pollutants, those for which only the level, not the location, of the emissions matters. This was followed shortly by an article by Montgomery (1972) that generalized the results for the more complicated case of non-uniformly mixed pollutants, those for which both the level and the location of emissions matters. These articles were instrumental in legitimizing the concept of emissions trading in the eyes of those theorists who tend to be distrustful of ad hoc arguments until the formal properties of the system are worked out.

## Describing the Evolution

### *Traditional Policy*

To understand how these general principles were implemented in the early years in the United States, it is important first to understand the policy environment that gave rise to the reform. Some knowledge of that policy framework helps not only to understand the forces for reform but also the shape of that reform once it began to happen.

U.S. air pollution policy was, and is, designed to ensure that people and ecosystems are protected from harmful levels of pollution. It docs this by promulgating ambient air quality standards that specify the permissible legal threshold for concentrations of pollutants in the ambient air and by establishing a process for reaching those standards.

The traditional approach for improving air quality typically involved the bureaucratic selection of desirable control technologies, using those technologies as the basis for specifying permissible emission limits, and forcing emitters to stay within those limits.

In the early 1970s, a group of experts from the academic community familiar with emerging literature on property rights suggested that it might be possible to improve upon this system by allowing firms to trade control responsibility among themselves by means of emissions trading. In this way, firms that could control relatively cheaply would voluntarily control more, selling the excess control to those that, for economic reasons, wanted to control less.

In an important sense, emissions trading changes the nature of the regulatory process. The burden of identifying the appropriate control strategies is shifted from the control authority to the polluter. As a result of the flexibility that becomes possible from this shift, many new control strategies can, in principle, emerge. Instead of the traditional focus on end-of-pipe control technologies, pollution prevention is placed on an equal footing by this program. All possible pollution reduction strategies can, for the first time, compete on a level playing field.

Emissions trading also allows more flexibility in the timing of control investments. Under emissions trading, facilities have the ability to time their expenditures so that they coincide with optimal capital replacement schedules and prevailing market conditions. Forcing every emitter to install control equipment at precisely the same time could cause much higher equipment purchase expenditures than would be the case with a schedule that spread deliveries out over a longer period. Demand would be less temporally concentrated with emissions trading.

Reform, however, has its own costs, and the existing policy was likely to persist unless it could be shown that the difference the new policy would make would be sufficiently large to justify the change. Making that case, of course, required empirical analysis that could begin to get at the magnitude of the potential cost savings involved. Were they large enough to justify the effort?

As discussed in more detail in the next chapter, the initial empirical analysis suggested that the command-and-control policy was very cost-ineffective. (Atkinson and Lewis 1974; Tietenberg 1974). Subsequent analyses involving several different pollutants in several different regions find that the initial empirical results were robust—the control costs from command-and-control allocations were estimated to exceed least cost allocations by a substantial margin (Seskin et al. 1983; Roach et al. 1981; Atkinson and Tietenberg 1982; Krupnick, Oates et al. 1983; Maloney and Yandle 1984; McGartland and Oates 1985). Though based on ex ante computer simulations that dealt more with caricatures of command-and-control than actual regulatory allocations, this literature offered the possibility that the increased flexibility made possible by the reform had the potential to meet the environmental targets with substantially lower control costs.

# The Evolution of Emissions Trading

## *The Offset Policy: The Problem Becomes the Solution*

The political opportunity to capitalize on that insight came in 1976. By 1976, it had become clear that a number of regions designated as "nonattainment" regions by the Clean Air Act would fail to attain the ambient air quality standards by the deadlines mandated in the act. Because the statute mandated improvements in air quality in these regions, further economic growth appeared to make the air worse, contrary to the intent of the statute. The Environmental Protection Agency (EPA) was faced with the unpleasant prospect of prohibiting many new businesses (those that would emit any of the pollutants responsible for nonattainment in that region) from entering these regions until the air quality came within the ambient standards.

Prohibiting economic growth as a means of resolving air quality problems was politically unpopular among governors, mayors, and many members of Congress. EPA had a potential revolution on its hands. At this point, they began to consider alternatives. Was it possible to solve the air quality problem while allowing further economic growth?

It was possible, as it turns out, and the means for achieving these apparently incompatible objectives involved the creation of an early form of emissions trading. Existing sources of pollution in the nonattainment area were encouraged to voluntarily reduce their emissions levels below their current legal requirements. EPA could then certify these excess reductions as emission reduction credits. Once certified by the control authority, these credits then became transferable to new sources that wished to enter the area; this created the supply of credits.

The demand for credits was created by new sources. New sources were allowed to enter nonattainment regions provided they acquired sufficient emissions reduction credits from other facilities in the region and that total regional emissions would be lower (not just the same) after entry than before. (This was accomplished by requiring new sources to secure credits for 120% of the emissions they would add; the extra 20% would be "retired" as an improvement in air quality.) Known as the offset policy, this approach not only allowed economic growth while improving air quality—it made economic growth the vehicle for improvement. It turned the problem on its head by making the problem part of the solution (Tietenberg 1985; Hahn and Hester 1989b).

### The U.S. Emissions Trading Program: Expanding the Scope

It wasn't long before the government began to expand the scope of the program by combining three new policies (the bubble policy, banking, and netting) with the offset policy into what became known as the Emissions Trading Program (ETP).

Whereas the offset policy allowed new sources to acquire credits from existing sources, the bubble policy allowed existing sources to acquire credits from each other. Banking simply allowed the created credits to be saved for subsequent use or sale in the future.

Netting allowed sources undergoing modifications or expansions to escape the burden of meeting the strict requirements imposed on new sources as long as any net increase in emissions (counting emissions reduction credits) fell below an established threshold. By netting out of this review, the facility would not only be exempted from meeting the strict new source technology requirements, but they also would be relieved of the need to secure offsets for the remaining emissions.

Like the offset policy, the emissions trading program was a credit program.[3] A firm could create credits by exceeding its legal reduction requirements and having the excess certified as an emissions reduction credit. In this program, the government not only was required to certify each reduction before it qualified for credit, but credit trades generally were approved by the control authority on a case-by-case basis.

## Getting the Lead Out: The Lead Phase-out Program

Following the introduction of the ETP, the government began applying the tradable permit approach more widely. One prominent use involved facilitating the regulatory process for getting lead out of gasoline (Hahn and Hester 1989a; Nussbaum 1992; Kerr and Maré 1997). In this case what was being controlled was not emissions per se, but an input: lead.

Lead was being added to gasoline by a relatively few refineries. Compared to monitoring and enforcing limits on the lead emissions from every gasoline-powered vehicle, it proved more administratively feasible to control the lead problem at the point of gasoline production.

In the mid-1980s, prior to the issuance of new, more stringent regulations on lead in gasoline, EPA announced the results of a cost–benefit analysis of their expected impact. The analysis concluded that the proposed .01 grams per leaded gallon standard would result in $36 billion (1983$) in benefits from reduced adverse health effects at an estimated cost of $2.6 billion to the refining industry.

Although these results suggested that regulation unquestionably was justified on efficiency grounds, EPA wanted to allow flexibility in how the deadlines were met without increasing the amount of lead used. While some refiners could meet early deadlines with ease, others could do so only with a significant increase in cost. Recognizing that meeting the goal did not require every refiner to make the transition at the same time, EPA initiated an artificial market in the rights to use lead in gasoline to provide additional flexibility in meeting the regulations.

Under this program, a fixed amount of lead rights (authorizing the use of a fixed amount of lead over the transition period) was allocated to the various refiners. Refiners who did not need their full share of authorized rights (due to early compliance) could sell their rights to other refiners.

Refiners had an incentive to eliminate the lead quickly because early reduction freed up rights for sale. Acquiring these credits made it possible for other refiners to comply with the deadlines even in the face of equipment failures or acts of God; fighting the deadlines in court, the traditional response, was unnecessary. Designed purely as a means of facilitating the transition to lead-free gasoline, the lead banking program ended as scheduled on December 31, 1987.

Several features of this program are noteworthy because they demonstrate the evolution of emissions trading into new territory. First, the lead program limited an input to emissions, not emissions directly. Second, the lead program was designed to eliminate a pollutant, not merely place an upper limit on its annual use. Third, it resulted in a much earlier phase-out of lead than traditionally would have been possible because of the inter-refinery flexibility it offered. The traditional approach—setting the deadline late enough to allow even refineries facing the most difficult compliance problems to meet it—would have resulted in a great deal more lead being injected into the air.[4] All of these features differentiate the lead program from the offset program.

## *Reducing Ozone-Depleting Chemicals*

Responding to the threat to the stratospheric ozone shield posed by the emission of ozone-depleting gases, 24 nations signed the Montreal Protocol during September 1988. According to this agreement, signatory nations were to restrict their production and consumption of the chief responsible gases to 50% of 1986 levels by June 30, 1998. Soon after the protocol was signed, new evidence suggested that it had not gone far enough; the damage was apparently increasing more rapidly than previously thought. In response, subsequent agreements called for the complete phase-out of halons and CFCs by the end of the twentieth century. Moreover, two other destructive chemicals—carbon tetrachloride and methyl chloroform—were added to the elimination schedule.

The United States chose to use a transferable permit system to implement its responsibilities under the protocols. On August 12, 1988, EPA issued regulations implementing a tradable permit system to achieve the targeted reductions. Under these initial regulations, all major U.S. producers and consumers of the controlled substances were allocated baseline production or consumption allowances using 1986 levels as the basis for their prorated share. Each producer and consumer was initially allowed 100% of this baseline allowance, with smaller allowances being granted after predefined deadlines.

These allowances were transferable within producer and consumer categories and allowances could be transferred across international borders to producers in other signatory nations if the transaction was approved by EPA and resulted in the appropriate adjustments in the buyer or seller allowances in their respective countries.[5]

Production allowances could be augmented by demonstrating the safe destruction of an equivalent amount of controlled substances by approved means. Some inter-pollutant trading was even possible within categories of pollutants. (The categories are defined so as to group pollutants with similar environmental effects.) All information on trades is confidential (known only to the traders and regulators), so it is difficult to know how effective this program has been. One estimate suggests that as of September 1993, the traded amount was roughly 10% of the total permits (Stavins and Hahn 1993).

Since the demand for these allowances was quite price inelastic, the supply restrictions imposed by this program increased revenue from their sale. By allocating allowances to the seven major domestic producers of CFCs and halons, EPA created sizable windfall profits (estimated to be in the billions of dollars) for those producers.

Those profits proved to be short-lived. A revenue-starved U.S. Congress seized the opportunity to impose a tax to soak up the rents created by the regulation-induced scarcity. The Revenue Reconciliation Act of 1989 included an excise tax that was imposed on all ozone-depleting chemicals sold or used by manufacturers, producers, or importers. The tax is imposed at the time the importer sells or uses the affected chemicals. It is computed by multiplying the chemical's weight by the base tax rate and the chemical's ozone-depletion factor. In addition to soaking up some of the regulation-induced scarcity rent, this tax provided an incentive to switch to less harmful (and therefore untaxed) substances. Over time, the tax rate grew high enough that it, not the allowances, was controlling the level of production and use. (In other words, at that point the demand for allowances was lower than the available supply.)

This application also contributed some new features to the evolution of emissions trading. It not only allowed international trading of allowances for the first time, but it involved the simultaneous application of emissions trading and taxes. (Prior to this application, taxes and allowances were considered substitutes, rather than complements.)

### Industrial Air Toxics: Emissions Averaging

The 1990 Clean Air Act Amendments required EPA to establish national emissions standards to control major industrial sources of toxic air pollution. In defining its approach, EPA has used emissions averaging as one of several ways to provide compliance leeway in these industry-by-industry standards. Emissions averaging allows facilities some flexibility in choosing the mix of sources used to meet a facility-wide cap; in essence, it involves intra-plant trading. For example, emissions averaging is permitted by national air toxics emissions standards for petroleum refining, synthetic organic chemical manufacturing, polymers and resins manufacturing, aluminum production, wood furniture manufacturing, printing and publishing, and a number of other sectors (Anderson 2001).[6]

### Tackling Acid Rain: The Sulfur Allowance Program

The most successful version of emissions trading to date has been its use in the U.S. approach for achieving further reductions in electric utility emissions contributing to acid rain. Under this innovative approach, allowances to emit sulfur oxides were allocated to plants; the number of authorized emissions was reduced in two phases to ensure a reduction of 10 million tons in emissions from 1980 levels by the year 2010 (Kete 1992; Rico 1995; Carlson et al. 2003; Ellerman et al. 2000; Ellerman 2003; Burtraw and Palmer 2004).

Perhaps the most interesting political aspect of this program was the role of trading in the passage of the acid rain bill. Though reductions of acid rain precursors had been sought with a succession of bills over the two decades of Clean Air Act legislation, none had become law. With the inclusion of an emissions trading program for sulfur dioxides in the bill, the compliance cost apparently was reduced sufficiently to make passage politically possible.

Sulfur dioxide allowances form the heart of this emissions trading program. The allowances are allocated to specified utilities on the basis of an allocation formula. Each allowance, which provides a limited authorization to emit one ton of sulfur, is defined for a specific calendar year, but unused allowances can be carried forward into future years. They are fully transferable not only among the affected sources but even to individuals who may wish to "retire" the allowances, thereby denying their use to legitimize emissions.

Emissions in this controlled sector cannot legally exceed the levels permitted by the allowances (allocated plus acquired). An annual year-end audit balances emissions with allowances. Utilities that emit more than authorized by their holdings of allowances must pay a per-ton penalty and are required to forfeit an equivalent number of tons the following year. The amount of the penalty was indexed to inflation from a starting value of $2,000 a ton.

This program has several innovative features, but in the interest of brevity only two are detailed here. The first important innovation in this program was ensuring the availability of allowances by instituting an auction market. Although allowances can be either transferred by private sale or the annual auction, the problem historically with the private sale route was that prices were confidential, so transactors were operating in the dark. Due to a lack of general knowledge not only about potential buyers and sellers but also about prices, transactions costs were high; before the Sulfur Allowance Program, emissions trading markets did not work very well.

EPA solved this problem by instituting an auction market run by the Chicago Board of Trade. In the negotiations, utilities fought the idea of an auction because they knew it would raise their costs significantly. Whereas under the traditional policy utilities would be given the allowances free of charge, under a conventional auction they would have had to buy these allowances at the full market price, a potentially significant additional financial burden.

To gain the advantages an auction offers for improving the efficiency of the market, while not imposing a rather large financial burden on utilities, EPA established what is now known as a zero revenue auction. Each year, EPA withholds from its allocation to utilities somewhat less than 3% of the allocated allowances and auctions them off. In the auction, these allowances are allocated to the highest bidders, with successful buyers paying their actual bid price (not a common market-clearing price). The proceeds from the sale of these allowances are refunded to the utilities from which the allowances were withheld on a proportional basis.

Private allowance holders other than utilities also may offer allowances for sale at these auctions. Potential sellers specify minimum acceptable prices. Once the withheld allowances have been disbursed, EPA then matches the highest remaining bids with the lowest minimum acceptable prices on the private offerings and matches buyers and sellers until all remaining bids are less than the remaining minimum acceptable prices. Although this auction design is not efficient, because it provides incentives for inefficient strategic behavior (Hausker 1992; Cason 1993), the degree of inefficiency is apparently small (Joskow et al. 1998; Ellerman et al. 2000).

A second innovation in this program is that it allows anyone to purchase allowances. This means that environmental groups or private citizens can buy them for the purpose of retiring them. Since retired allowances represent authorized emissions that are never emitted, they result in cleaner air.

## RECLAIM: The States Take the Initiative

All of the previous U.S. programs were initiated by the federal government. The states primarily were involved as the enforcement arm of the federal programs. In 1994, states became initiators as well as enforcers. Faced with the need to reduce pollutant concentrations considerably in order to come into compliance with the ambient standards, many states chose to use trading programs as a means of facilitating drastic reductions in emissions.

One of the most ambitious of these programs, and the focus of this section, is California's Regional Clean Air Incentives Market (RECLAIM), established by the South Coast Air Quality Management District, the district responsible for the greater Los Angeles area (Goldenberg 1993; Fromm and Hansjurgens 1996; Hall and Walton 1996; Johnson and Pekelney 1996; Harrison 2004).

Under RECLAIM, each of the almost 400 participating industrial polluters are allocated an annual pollution limit for nitrogen oxides and sulfur. This limit decreases by 5% to 8% each year for a decade. Polluters are allowed great flexibility in meeting these limits, including purchasing RECLAIM tradable credits (RTC) from other firms that have controlled more than their legal requirements.

In part to ensure that the program actually could be implemented, the initial authorization for emission limits was generous. As a result, in the early years of the RECLAIM program (1993–1999) most companies had an excess number of RTC credits because the initial allocation exceeded actual emissions rates.

During the summer of 2000, problems with California's electricity deregulation program spilled over into the RECLAIM market, resulting in a sharp and sudden increase in credit prices (Joskow and Kahn 2002). Specifically, "the average price of $NO_x$ RTCs for compliance year 2000, traded in the year 2000, increased sharply to over $45,000 per ton compared to the average price of $4,284 per ton traded in 1999" (SCAQMD 2001).

The district responded by temporarily pulling power plants out of the program and instituting a mitigation fee as a safety valve. Under the mitigation fee alternative, power-producing facilities contribute $7.50 per credit for any emissions exceeding their allocation. The collected funds are invested in emission reduction projects.

Aside from the fact that this was the first emissions trading program initiated by a state, two other features of this program are of interest. First, trading within this program was restricted spatially by the creation of trading zones for the coastal and interior regions. Furthermore, RECLAIM is the only program to face such significant price increases. As will become clear in subsequent chapters, the experience with both of these features provides some valuable lessons for program design.

## The $NO_x$ Budget Program

The next phase in the evolution involved states getting together to cooperate in the development and initiation of a regional cap-and-trade program. States in the Northeastern United States collaborated via the Ozone Transport Commission (OTC) to achieve ozone-season $NO_x$ reductions in several phases.[7] In Phase I, sources were required to reduce their annual rates of $NO_x$ emissions to meet Reasonably Available Control Technology requirements. In Phase II, states initiated a cap-and-trade program, named the OTC $NO_x$ Budget Program (OTCNBP), to achieve additional reductions during the ozone season.

In 2003, the OTCNBP was replaced in Phase III by the geographically more inclusive[8] $NO_x$ Budget Trading Program (NBP). While the Phase III NBP was initiated federally, it essentially ratified and extended the initiative that had already been undertaken by the Northeastern states. Although attainment of the federal ozone ambient standard was the motivation, it was the OTC states that decided to implement a seasonal cap-and-trade program starting in 1999. (Emissions are capped during the ozone season from May 1 to September 30.) It was an outgrowth of the discussions within the OTC set up by the 1990 Clean Air Amendments to facilitate interstate cooperation in the Northeast.

As a result of this earlier state-initiated program, states in the OTC were able to comply with the NBP in 2003, and many other states across the East and Midwest began to reduce emissions in 2004. Twenty-one states and the District of Columbia are participating or will participate in the future.

Based on data reported to EPA, nearly 2,600 affected and operating units are in the NBP states, including states that joined the program in 2004. About 1,000 of these units are in OTC states that complied in 2003, while another 1,600 units are in non-OTC states.

The most unique feature of this program is its focus on seasonal emissions. Whereas most emissions trading programs focus on reducing annual emissions, this program, recognizing the important seasonal component in the formation of ozone in the Northeast, focuses only on those emissions that will directly contribute to the problem.

## The Chicago Emissions Reduction Market System

The Chicago area, which was a severe nonattainment area for ozone, was required to come into compliance with the ozone ambient standard by 2007. Concluding that meeting the standard would require substantial reductions in volatile organic matter (VOM) emissions, Illinois adopted the Emissions Reduction Market System (ERMS), an emissions trading program that would reduce overall VOM emissions in the Chicago area.

Similar to the $NO_x$ budget program (because VOM emissions contribute to ozone formation as well), the ERMS program operates from May 1 through September 30. The program allows trading among participating sources to meet a reduced cap on their overall VOM emissions. Each participant is given a number of allotment trading units, corresponding to an overall reduction of 12%, according to what they emitted in previous years.

This was the first program to attempt to use emissions trading to control volatile organic materials. A proposed earlier effort to use a cap-and-trade mechanism in the Los Angeles area to limit VOM emissions had been discussed but not implemented. Implementing a cap-and-trade system for VOM is especially complex because VOM sources not only include a wide array of very different types of emitters, many of them small, but volatile organics represent a class of pollutants, not a single pollutant. To make matters even more complex, VOM contains substances that have been designated as hazardous air pollutants due to the danger they pose to human health. Clearly not all VOM reductions would have the same environmental impact should trading change the mix.

To date, the evidence suggests that this program was superfluous and therefore ineffective. While ex post emissions were far below the cap, findings of unexpectedly large banks, startling permit expirations, and low prices of tradable permits all provide evidence of an ineffective market. Kosobud et al. (2004) find that the market as designed had been constrained from reaching its objectives by the continuance and extension of an underlying layer of traditional regulation and to a lesser extent by over-allotment of tradable permits. The regulations, not emissions trading, largely were responsible for the reductions, and the responses they mandated left little room for emissions trading to play any role.

## Emissions Trading in the Kyoto Protocol on Climate Change

The 1992 United Nations Framework Convention on Climate Change (UNFCCC) recognized the principle of the global cost-effectiveness of emissions reduction and opened the way for flexibility in how greenhouse gas targets would be met. Because this early agreement did not fix a binding emissions target for any country, however, the need to invest in emissions reduction at home or abroad was not pressing.

In December 1997, however, industrial countries and countries with economies in transition (primarily the former Soviet Republics) agreed to legally binding emissions targets at the Kyoto Conference and negotiated a legal framework as a protocol to the UNFCCC—the Kyoto Protocol. This protocol became effective in February 2005 once Russia's ratification put the Kyoto Protocol over the 55% threshold.[9]

The Kyoto Protocol defines a five-year commitment period (2008–2012) for meeting the individual country emissions targets, called "assigned amount obligations," set out in Annex B of the protocol. Quantified country targets are defined by multiplying the country's 1990 emissions level by a reduction factor and multiplying that number by five (to cover the five-year commitment period). Collectively, if fulfilled, these targets would represent a 5% reduction in annual average emissions below 1990 levels from this group. The actual compliance target is defined as a weighted average of six greenhouse gases: carbon dioxide, methane, nitrous oxide, HFCs, PFCs, and sulfur hexafluoride. Defining the target in terms of this multi-gas index, rather than only carbon dioxide, has been estimated to reduce compliance costs by some 22% (Reilly et al. 2002).

The Kyoto Protocol authorizes three cooperative implementation mechanisms that involve tradable permits. These include Emission Trading, Joint Implementation, and the Clean Development Mechanism.

- Emissions Trading allows trading as a means of fulfilling the national quotas established by the Kyoto Protocol among countries listed in Annex B of the protocol, primarily industrialized nations and economics in transition. Although in principle trading of emissions reductions could occur among private parties as well as countries, the enabling article, Article 17, is silent on this point.
- Under Joint Implementation, Annex B parties can receive emissions reduction credit when they help to finance specific projects that reduce net emissions in another Annex B party country. This project-based program is designed to exploit opportunities in Annex B countries that have not yet become fully eligible to engage in the Emissions Trading Program described above.
- The Clean Development Mechanism enables Annex B parties to finance emissions reduction projects in non-Annex B parties (primarily developing countries) and to receive certified emission reductions (CERs) for doing so. These CERs could then be used, along with in-country reductions, to fulfill assigned amount obligations.

## The European Union Emissions Trading Scheme

The largest and most important emissions trading program has been developed by the European Union to facilitate implementation of the Kyoto Protocol

(Kruger and Pizer 2004). The EU Emissions Trading Scheme (EU ETS) applies to 25 countries, including the 10 accession countries, most of which are former members of the Soviet bloc.[10] The first phase, from 2005 through 2007, is considered to be a trial phase. The second phase coincides with the first Kyoto commitment period, beginning in 2008 and continuing through 2012. Subsequent negotiations will specify the details of future phases.

Initially, the program will cover only carbon dioxide emissions from four broad sectors: iron and steel, minerals, energy, and pulp and paper. All installations in these sectors larger than established thresholds are included in the program. Some 12,000 installations are expected to be included, the largest number of sources covered by any program.

Individual countries determine the initial allocation using a two-step process. First, they must decide how much of the predefined national cap should be allocated to each of these sectors and second, how much of each sector cap should be allocated to each of the installations in that sector. Making these initial allocations has turned out to be a highly controversial process, in part because competitors in different European countries could end up with quite different allocations (and therefore different costs of compliance).

Though this allocation scheme provides installations with the permits for free, auctions of permits may be held in the future. Countries will be allowed the option of auctioning up to 5% of allowances in the first phase of the program and up to 10% in the second phase. Participating countries can use credits acquired from outside the European Union (via the Joint Implementation or Clean Development Mechanisms) to meet their obligations under the emissions trading system.

A couple of other features differentiate this system from U.S. programs. In the EU ETS, new plants in many cases are granted gratis permits, whereas in the U.S. system new firms usually have to buy any permits they need. In addition, plant closures tend to lead to the forfeiture of permits more often in the EU system than the U.S. system (Åhman et al. 2005).

## Controlling Particulates in Santiago, Chile

Emissions trading is being implemented in developing counties as well (Montero et al. 2002). In March 1992, a program to control total suspended particulate emissions from stationary sources was established in Santiago, Chile. Sources registered and operating at that time received grandfathered permits, while new sources, which received no permits, must cover or offset all their emissions by buying permits from existing sources.[11]

These permits are denominated not in terms of a source's actual emissions but in terms of its "emissions capacity," which is equal to the maximum emissions that the source could potentially emit in a given period of time.[12] The regulator annually reconciles the estimated emissions capacity with the number of capacity permits held by each source. Firms with actual capacity that falls below the authorized capacity can sell the excess permits, while firms with too few permits can purchase them from other sources. Although permits are traded at a 1:1 ratio, all trades require approval by the regulatory agency.[13]

The Santiago program occupies a significant place in the evolution of emissions trading because it provides experience about the effectiveness of these programs in the environment of a developing country. As such, it provides some important insights into the infrastructure requirements for emissions trading, a subject discussed in more detail in subsequent chapters.

### Clean Air Interstate Rule

New, national health-based air quality standards in the United States for ozone and particulate matter (PM2.5) require substantial reductions beyond the existing regulations. On March 10, 2005, EPA announced the Clean Air Interstate Rule, a rule that covers 28 eastern states and the District of Columbia.

States must achieve the required emissions reductions using one of two compliance options: (1) meeting the state's emissions budget by requiring power plants to participate in an EPA-administered interstate cap-and-trade system that caps emissions in two stages; or (2) meeting an individual state emissions budget through measures of the state's choosing.

This reliance on a cap-and-trade approach to securing these reductions is a large step in the evolution of cap-and-trade programs. Normally, state compliance is determined through EPA oversight and approval of state implementation plans. In this case, state actions are deemed adequate if the federally designed model rule of emissions trading is adopted. This EPA dependence on the mechanisms set up by the Clean Air Act Amendments of 1990 would have been unimaginable 20 years ago and may represent the clearest example of the extent to which emissions trading has replaced conventional source-by-source regulation.[14]

# The Evolution of Design Features

The evolution of emissions trading over time has resulted in considerable change in some of the programmatic design elements.

## Permit Denominations

The original ETP was based on a system of credits that typically were denominated in terms of a pollutant flow such as tons per year. The newer programs are based on allowances defined in discrete terms (e.g., tons rather than tons/year). While the former conferred a continuing entitlement to a flow, the latter provides a one-time entitlement to emit a specific quantity.

Though seemingly a small change, in fact the opportunity for discrete credits has proved to be quite important. One of the original criteria used by EPA for approving credits was that the emissions reduction supporting them should be permanent. Many useful strategies to reduce emissions, such as meeting a deadline early, produce temporary, rather than permanent, reductions. (As noted above, the ability to set an earlier deadline in the Lead Phase-out Program was made possible by the flexibility inherent in an allowance program.) Because allowance programs encourage discrete as well as permanent flow reductions, they facilitate these additional cost reductions with no adverse impact for the environment.

Authorizing future emissions in an allowance system requires the issuance of future allowances. In general, this is done well in advance of the applicable dates according to specific schedules so emitters have reasonable security for pollution control investment planning. Allocating allowances in advance has also facilitated the development of futures markets.[15]

Another major difference is that credits generally have to be certified as excess in advance of any trade by the control authority. In contrast, allowances can be freely traded without any certification step. Since compliance with allowances is ascertained by an end-of-period comparison between actual emissions and surrendered allowances, no certification step is necessary.

## Seasonality

For some pollutants, such as ozone, there is a strong seasonal element in the relationship between emissions patterns of precursor pollutants and the resulting ozone concentrations. Another significant characteristic of the evolution of emissions trading is the tendency to tailor programs to take this relationship into account by targeting emissions that occur during the ozone-formation period. This kind of targeting occurs in both the NBP and the Chicago VOM program.

## Baseline

Credit trading, the approach taken in the bubble and offset policies, allows emissions reductions above and beyond legal requirements to be certified as tradable credits. The baseline for credits is provided by traditional technology-based standards. Credit trading presumes the preexistence of these standards and it provides a more flexible means of achieving the source-specific goals that the source-based standards were designed to achieve.

Allowance trading, used in the Acid Rain Program and RECLAIM in California, assigns a pre-specified number of allowances to polluters—a number that may or may not have anything to do with the historical, technology-based standards. Typically, the number of issued allowances declines over time, and in most cases the magnitude of the aggregate reductions implied by the allowance allocations exceed those achievable by standards based on currently known technologies.

Despite their apparent similarity, the difference between credit and allowance-based trading baselines should not be overlooked. Credit trading depends upon the existence of a previously determined set of regulatory standards; allowance trading does not. Once the aggregate number of allowances is defined, they can, in principle, be allocated among sources in many different ways. The practical implication is that allowances can be used even in circumstances where a technology-based baseline either has not, or cannot, be established or where the control authority wished to allocate permits in some way other than historical, technology-based standards.[16]

## Caps

The tendency for emissions trading systems to move over time from credit systems, such as the U.S. ETP, to allowance systems, such as the NBP and sulfur allowance systems, has another important implication. Allowance systems set a cap on aggregate emissions that is not eroded by the entrance of new emitters. This limit on aggregate emissions is not shared by traditional, technology-based, source-specific emissions standards or, in the absence of other constraints, by an emissions credit system that is linked to technology-based standards. Because emissions standards are source-specific, they exert no control over the aggregate amount of emissions from all sources. As the number of sources increases, the aggregate level of emissions increases. As a consequence, credit trading, which is based on these source-specific standards, will allow aggregate emissions increases unless additional constraints are built into the system.

In the United States, the additional constraint in the ETP was mandating offsets in nonattainment areas. Requiring that all new emitters secure sufficient credits from existing emitters so that air quality would improve as a result of their entering the area provides a cap on aggregate emissions. Since no such constraint was mandatory in attainment areas, credit trading provided no protection from emissions increases as the number of sources increased in those areas.

## Shifting the Payoff

The demonstration that traditional regulatory policy was not cost-effective had two mirror-image implications. It either implied that the same air quality could be achieved at lower cost or that better air quality could be achieved at the same cost. While the earlier programs were designed to exploit the first implication, the later programs attempted to produce better air quality and lower costs.[17]

Trading programs were used to produce better air quality in many ways. The lower costs offered by trading were used in initial negotiations to secure somewhat more stringent pollution control targets (as in the Acid Rain Program and RECLAIM) or earlier deadlines (as in the Lead Phase-out Program). Offset ratios for trades in nonattainment areas were set at a ratio greater than 1, implying that a portion of each acquisition would go for better air quality. Environmental groups are allowed to purchase and retire allowances (Acid Rain Program).

This shift toward sharing the benefits between environmental improvement and cost reduction has had two consequences. The cost savings are smaller than they would have been without this benefit sharing, but the public support, and particularly the support from environmental organizations, probably has been increased a great deal.

### Substituting for vs. Complementing Traditional Regulation

The earliest use of the tradable permit concept, the ETP, overlaid credit trading on an existing regulatory regime and was designed to facilitate implementation of that program. Trading baselines were determined on the basis of previously determined, technology-based standards and created credits could not be used to satisfy all of these standards. For some, the requisite technology had to be installed.

More recent programs, such as the Acid Rain and RECLAIM programs, replace, rather than complement, traditional regulation. Allowance allocations for these programs were not based on preexisting, technology-based standards. In the case of RECLAIM, the decadal declines were sufficiently large that the control authority (the South Coast Air Quality Management District) could not have based allowances on predetermined standards even if they had been inclined to do so. Defining a complete set of technologies that offered the necessary environmental improvement and yet were feasible in both an economic and engineering sense, proved a formidable, if not impossible, challenge. Traditional regulation was incapable of providing the huge degree of reduction within the deadlines required by the Clean Air Act.

The solution was to define a set of allowances that would meet the environmental objectives, leaving the choice of methods for living within the constraints imposed by those allowances up to the sources covered by the regulations. This approach fundamentally changed the nature of the control process. The historical approach involved making the control authority responsible not only for defining the environmental objectives and performing the monitoring and enforcement activities necessary to ensure compliance with those objectives, but it also was assigned the responsibility for defining the best means for reaching those objectives. Emissions trading transfers the last of these responsibilities to the private sector, while retaining for the public sector both the responsibility for defining the environmental target and monitoring and enforcing compliance with it.

The other major change, seen in the Clean Air Interstate Rule, is substituting a pre-approved cap-and-trade program for state-designed approaches that have to be included in the state implementation plan and ultimately approved on a case-by-case basis by EPA.

## An Overview of the Book

Emissions trading certainly did not command an immediate constituency and building one was not easy. In the early days, even industrial sources, the most natural constituents in light of their potential cost savings and the flexibility of the program, were far from unanimous in their enthusiasm.

To some extent, industrial sources feared that this flexibility entailed greater risk. When an EPA-recommended technology failed to live up to standards, the source could claim that it had lived up to its responsibilities, so was blameless. When the control mix was up to the source, however, it would lose this defense. Similarly, by reducing emissions more than was required by law in order to gain emissions reduction credits, plants would alert control authorities to the fact that additional control was possible. Should control authorities use this information to revise the control baseline upward, similar plants under the same ownership would be adversely affected by the creation of the credits. Industrialists feared that they could end up losing more in the long run than they gained in the short run.

Industrialists were not the only constituency to have concerns about emissions trading. State authorities, which in the early U.S. program would bear the brunt of implementation, feared that the new programs would be more difficult to administer and saw them as a threatening departure from their comfortable, customary way of doing business. Environmentalists feared, and no doubt some industrialists hoped, that the program would open a large number of loopholes, leaving a legacy of reduced compliance.

Despite this opposition, emissions trading has now gained a firm foothold in environmental policy and is likely to continue to expand into new arenas for some time. Since so many ideas for policy reform have failed to come to fruition, it is natural to ask what sacrifices were made to place this type of program on the books. Compromise is, after all, the essence of most successful reforms. To what extent were the stated goals of emissions trading programs— increased cost effectiveness and increased speed of compliance—compromised as the price of initiating and maintaining the program?

Equally as important is the question of whether the expectations emanating from theory were unrealistically high when judged in retrospect by actual experience. Were costs minimized? Was the level of environmental protection effective and efficient? Was technical innovation in abatement technologies promoted?

The evidence on which the evaluation in this book is based is drawn from three different sources:

- Economic theory, the first source, is used to derive both market and optimality conditions for various market conditions and pollutant characteristics. It also is used to formalize the incentives present in various permit designs and show whether or not those incentives are compatible with the programmatic objectives.

- Ex ante computer simulations, the second source, flesh out the bare bones of theory. They provide a foundation for an investigation of the magnitude of the control cost differences between the command-and-control and emissions trading allocations, as well as for an identification of the sources of those differences. By incorporating the specific meteorological and source configuration characteristics unique to each region studied, computer simulations bring the general results of theory into sharper focus.

- The final source, actual emissions trading procedures and transactions, allows us to study in some detail how the programs have worked in practice.

The remainder of the book is divided into three parts. The first part, consisting of Chapters 2 and 3, lays the groundwork for the detailed analysis of specific program attributes that follow. Chapter 2 develops the theory behind the emissions trading program, and Chapter 3 estimates the magnitude of potential cost savings revealed by ex ante simulations and the actual results revealed by ex post evaluations.

The next portion of the book is concerned with evaluating the manner in which emissions trading programs have coped with a number of practical implementation problems. Chapters 4 and 5 open this section by examining the spatial and temporal dimensions of emissions trading. Since the initial allocation turns out to be one of the most politically volatile and important aspects of emissions trading, chapter 6 explores this issue in depth. The last two chapters in this section cover the potential for and consequences of market power (Chapter 7) and issues related to the monitoring and enforcement of emissions trading systems (Chapter 8).

The final chapter of the book, Chapter 9, weaves together the insights gained from the individual topic-by-topic examinations of the programs to form a comprehensive evaluation. This final chapter characterizes the state of the art and extracts lessons that might be drawn from this review.

# Notes

1. For an extended bibliography of works that describe these various applications see: http://www.colby.edu/~thtieten/trade.html.

2. Setting an efficient goal would, of course, require that information. In practice, governments frequently have been satisfied with "reasonable" goals.

3. This national "credit" version of emissions trading subsequently has been adopted as a model for some state programs. For example, Michigan implemented its own version of this program on March 16, 1996. See that program's Web site at: http://www.michigan.gov/deq/0,1607,7-135-3310_4103_4194-10617—,00.html.

4. Early reductions were especially important in this case because, as the cost–benefit analysis persuasively demonstrated, the health consequences of ambient lead were severe, particularly on children.

5. Note that this approach does not require that both trading countries have implemented a transferable permit system. It does require both countries to adjust their production and consumption quotas assigned under the protocols to ensure that overall global limits on production and consumption are not affected by the trades. The European Union also implemented a tradable permits scheme for ozone-depleting chemicals. See Council Regulation (EEC) No. 594/91 of March 4, 1991, on substances that deplete the ozone layer, Official Journal of the European Communities, 14.3.91.

6. A number of smaller programs also were introduced during this time, including reformulated gasoline, heavy-duty truck emissions averaging, and the hazardous air pollutant early reduction program, among others. For a description of these programs, see Anderson (2001, Chapter 6).

7. The OTC states that participated in this initial trading program were Connecticut, Delaware, Maryland, Massachusetts, New Hampshire, New Jersey, New York, Pennsylvania, Rhode Island, and the District of Columbia.

8. The OTC states were joined in the $NO_x$ State Implementation Plan call by Alabama, Illinois, Indiana, Kentucky, Michigan, North Carolina, Ohio, South Carolina, Tennessee, Virginia, and West Virginia on May 31, 2004.

9. The treaty was scheduled to go into effect only if two conditions were met: (1) at least 55 nations had to ratify the protocol; and (2) the ratifying nations had to account for at least 55% of total carbon equivalent emissions. The first condition proved easy to meet, but due to the failure of the United States, the largest emitter, to ratify the protocol, the second condition was met only when Russia ratified.

10. Details on this program can be obtained from its Web site at: http://europa.eu.int/comm/environment/climat/emission.htm

11. For new sources entering the program during or after 1998, 120% must be offset. This means new sources are required to buy 20% more emissions capacity than they would need. Because the extra 20% remains unused, it is a source of improved air quality.

12. Formally, it was defined as the product of emissions concentration (in $mg/m^3$) and maximum flow rate (in $m^3/hour$) of the gas exiting the source's stack.

13. Even a trade between two existing sources sharing common ownership required regulatory approval.

14. I am indebted to Denny Ellerman (MIT) for pointing this out to me.

15. In a futures market, buyers can acquire allowances for a specified future date (say five years hence) at a current market-determined price. Enabling sources to plan ahead by fixing the future allowance price now reduces the risk of price fluctuations and uncertainty about the effects of abatement investments.

16. This was an important source of flexibility for the initial allocation under the Sulfur Allowance Program, as Raymond (2003) points out.

17. One interesting analysis examines the cost and emissions savings from implementing an emissions trading system for light-duty vehicles in California. In that study, Kling (1994a) finds that although the cost savings from implementing an emissions trading program (holding emissions constant) would be modest (on the order of 1% to 10%), the emissions savings possibilities (holding costs constant) would be much larger (ranging from 7% to 65%).

# 2

# The Conceptual Framework

Why has emissions trading become a key component of environmental policy reform? To answer this question, as well as to provide a basis for evaluating how successful the reforms have been, some notion of an optimal allocation of control responsibility must be defined. In this chapter, the theory behind cost-effectiveness, the main basis for current regulation, is developed and used as one of the benchmarks against which the existing system can be measured.[1]

## The Regulatory Dilemma

Two principal types of participants are crucial in the process of regulating the amount of pollution in the air. While the regulatory authority has the statutory responsibility for ensuring acceptable air quality, those managing the sources of the pollutant (such as industries, automobiles, power plants, etc.) must ultimately take the actions that will reduce pollution sufficiently to meet specified goals. The key to successful regulation is to design programs that harmonize the efforts of these two groups.

## The Cost-Effectiveness Framework

The main responsibilities of the regulator are to decide how to allocate control responsibility among the sources, to design the regulations implementing those decisions, to monitor compliance with the resulting regulations, and to bring enforcement actions against those found not in compliance. This is a challenging responsibility if for no other reason than the very large number of sources that must be regulated.

As noted in the first chapter, the U.S. regulatory authority traditionally pursued its mandate by establishing separate emission standards for each point of discharge from major sources of the pollutant. Following the literature on this subject, this means of distributing control responsibility among points of discharge henceforth will be called the command-and-control approach. Since each industrial plant typically contains several pollutant discharge points, each with its unique emissions standard, the amount of information that the control authority needs to define cost-effective standards is staggering. Typically, the amount of information available to the regulatory authority when the allocations are made is insufficient.

In some ways, the managers of the plants emitting pollution are in exactly the opposite position. Because each plant manager typically would know (or know how to find out) the unique array of possible control techniques that are most suited to his or her operation, as well as the associated costs and reliability of these techniques, the quality of information at this level of decisionmaking is very good. Plant managers generally would have an excellent feel for the control technologies that would produce the most cost-effective emission reductions at their plants.

Unfortunately, however, plant managers lack the incentive to act on this information in a manner consistent with cost-effective emissions reduction. Since any unilateral increase in cost incurred by individual plants faced with competition, either from existing or potential rivals, could weaken their competitive position, plants would seek to minimize their own costs using any means at their disposal. Possible means include overstating costs to the control authority or to the legislature in hopes of being allocated a weak standard or seeking an exemption from the courts on grounds of affordability or technological infeasibility.

The fundamental problem with the command-and-control approach is a mismatch between capabilities and responsibilities. Those with the incentive to allocate the control responsibility cost effectively—the control authorities—have too little information available to them to accomplish this objective. Those with the best information on the cost-effective choices—the plant managers—have no incentive either to voluntarily accept their cost-effective responsibility or to transmit unbiased cost information to the control authority so it can make a cost-effective assignment. Plant managers have an incentive to accept as little control responsibility as possible in order to maintain or strengthen their competitive positions.

In this policy environment, it is not surprising that the command-and-control allocation is not, and by itself could not become, cost-effective. What may be surprising in light of the complexity of the task is that cost-effectiveness is not an unreasonable objective for other approaches.

# Cost-Effective Permit Markets

The reason emissions trading can result, in principle, in a cost-effective allocation is quite straightforward. Plants typically have very different costs of controlling emissions. When emissions permits are transferable, those plants that can control most cheaply find it in their interest to control a higher percentage of their emissions because they can sell the excess permits. Buyers can be found whenever it is cheaper to buy permits for use at a particular plant than to install more control equipment. Whenever an allocation of control responsibility is not cost-effective, further opportunities for trade exist. When all such opportunities have been fully exploited, the allocation would be cost-effective.

In principle, emissions trading solves the problems of information and incentive that are posed by the command-and-control approach by allowing each participant to play that role it plays best. Regulators ensure that sources have the proper incentives by setting the pollution targets and enforcing compliance with them. By exploiting the flexibility inherent in emissions trading to lower their own costs within the boundaries established by the control authority, sources collectively lower the total costs incurred by all sources. In principle, the incentives generated by self-interest in this case are compatible with cost-effectiveness.

Though these general principles hold for all pollutants, some of the implementation details (such as the design of the permit) depend crucially on the nature of the pollutant being regulated. Three different classes of pollutants are considered in this chapter. The definition of the cost-effective allocation as well as the design of the emissions trading system that is compatible with that allocation varies among these pollutant classes. The characteristic that distinguishes any one of these pollutant classes from another is the relationship between individual source emissions and the pollution target.

## Uniformly Mixed Assimilative Pollutants[2]

The first, and in a number of ways the simplest, class of pollutants to control conventionally is named uniformly mixed assimilative pollutants. In the case of assimilative pollutants, the capacity of the environment to absorb them is sufficiently large, relative to their rate of emission, so that the pollution level in any year is independent of the amount emitted in previous years. Simply put, assimilative pollutants do not accumulate over time.

In the case of uniformly mixed pollutants, the ambient concentration depends on the total amount of emissions but not on the distribution of these emissions among the various sources. For this class of pollutants, all possible distributions of the control responsibility within an airshed yielding the same total emissions will produce approximately the same effect on the pollution target. Greenhouse gases provide one example of a type of pollution that fits this description, since their contribution to climate change is not thought to be sensitive as to where they are emitted in the atmosphere.

Each of these characteristics (assimilation and uniform mixing) limits the complexity of a cost-effective permit system. The former allows the control process to ignore the difficult problem of pollutant accumulation, while the latter eliminates the need to worry about the location of the sources in designing the control policy—a significant advantage.

Symbolically, the relationship between source emissions and the pollution target for a uniformly mixed assimilative pollutant can be written as

$$A = a + b \sum_{j=1}^{J} (\bar{e_j} - r_j) \tag{2-1}$$

where A is the steady-state level of pollution in a year, $\bar{e_j}$ is the steady-state emission rate of the $j$th source that would prevail if the source failed to control any pollution at all (hereafter referred to as the uncontrolled emission rate), $r_j$ is the amount of emissions reduction achieved by the $j$th source, J is the total number of sources, and both "a" and "b" are parameters. Typically, the "a" parameter is used to represent background pollution (from natural sources or sources that for one reason or another are not regulated), and the "b" parameter is simply a constant of proportionality.

Cost effectiveness in this context is defined as that allocation of emissions levels among the $J$ sources that meets the pollution target (designated $\bar{A}$) at minimum cost. Let $C_j(r_j)$ be the continuous cost function that represents the minimum cost to the source of achieving any level of emissions reduction $(r_j)$. Generally, as $r_j$ increases the marginal cost of control can also be presumed to increase.

Mathematically, the cost-effective allocation is the solution to the following minimization problem:

$$\min_{r_j} \sum_{j=1}^{J} C_j(r_j) \tag{2-2}$$

subject to                          $a + b \sum_{j=1}^{J} (\bar{e_j} - r_j) \leq \bar{A}$

and                          $r_j \geq 0 \qquad j = 1,...,J$

The necessary and sufficient conditions (the Kuhn-Tucker conditions) for an allocation of control responsibility among the sources (a $J$-dimensional-vector) to be cost effective are

$$a + b\sum_{j=1}^{J}(\overline{e_j} - r_j) \leq \overline{A} \qquad (2\text{-}3)$$

$$r_j\left[\frac{\partial C_j(r_j)}{\partial r_j} - \lambda b\right] \geq 0 \qquad j=1,...,J \qquad (2\text{-}4)$$

$$a + b\sum_{j=1}^{J}(\overline{e_j} - r_j) \leq \overline{A} \qquad (2\text{-}5)$$

$$\lambda\left[a + b\sum_{j=1}^{J}(e_j - r_j) \leq \overline{A}\right] = 0 \qquad (2\text{-}6)$$

$$r_j \geq 0; \quad \lambda \geq 0 \qquad j=1,...,J \qquad (2\text{-}7)$$

The $\lambda$ variable, known as the Lagrange multiplier, is introduced as a convenient means of solving this type of constrained optimization problem. It has a simple economic interpretation—it is the amount of control cost at the margin that could be saved if the environmental quality constraint, $\overline{A}$, were relaxed by one unit. It is a measure of the marginal difficulty of meeting the $\overline{A}$ standard.

These equations have a straightforward interpretation. In a cost-effective allocation of a uniformly mixed assimilative pollutant, the marginal cost of control for each source would be equal to the same constant ($\lambda b$). This implies that the marginal cost of control would be the same for all sources. As we shall see, this is a highly significant finding for the choice of policy instruments.

Other properties of the cost-effective allocation can also be extracted from these conditions. If, in a cost-effective allocation, the marginal cost of controlling the first unit of emissions reduction for that source is higher than $\lambda b$ and, hence, higher than the marginal costs of all sources controlling non-zero amounts, that particular source is not assigned any control responsibility (that is, $r_j = 0$). It must also be the case that $\lambda > 0$ as long as some control is needed. The condition $\lambda = 0$ simply implies that the uncontrolled emissions satisfy the environmental constraint; no control is necessary.

The cost-effective permit design for a uniformly mixed assimilative pollutant is called an emissions permit. An emissions permit is defined in terms of allowable emissions. To initiate an emissions-permit system, the control authority must define the aggregate amount of emissions $N$, which will be allowed from all emitters. This is calculated as

$$N = \sum_{j=1}^{J}(\overline{e_j} - r_j) = \frac{\overline{A} - a}{b} \qquad (2\text{-}8)$$

since this is the level of allowed emissions that will cause the environmental quality standard to be met with equality.

For some pollutants, both the "a" and/or "b" parameters change seasonally, implying that the number of permits should change seasonally as well. When the permit system reflects these seasonal effects, different marginal costs of control can prevail in different seasons. (Note the relevance of this insight for both the NBP and Chicago ERMS programs described in the previous chapter.)

Once the aggregate emissions reduction, $N$, is defined, it provides the basis for a set of emissions permits $q^o$. Suppose that each source has some initial endowment of these permits ($q_j^o$). Across all sources, this initial endowment must be equal to the number of allowable permits, that is,

$$\sum_{j=1}^{J} q_j^o = N$$

to ensure compliance with $\overline{A}$.

Faced with the need to choose a nonnegative level of control, the $j$th source's choice can be characterized as

$$\min_{r_j} C_j(r_j) + P(\overline{e_j} - r_j - q_j^o) \qquad (2\text{-}9)$$

where  is the price the source would pay for an acquired permit or receive for a permit sold to another source.

The solution for the set of all sources is

$$\frac{\partial C(r_j)}{\partial r_j} - P \geq 0 \qquad j = 1,...,J \qquad (2\text{-}10)$$

$$r_j \left[ \frac{\partial C(r_j)}{\partial r_j} - P \right] = 0 \qquad j = 1,...,J \qquad (2\text{-}11)$$

$$r_j \geq 0 \qquad j = 1,...,J \qquad (2\text{-}12)$$

This market solution can now be compared with the cost-effective solution. According to equation 2-8, the number of permits issued is compatible with $\overline{A}$, so as long as all firms comply with the terms of their permits, the environmental target would be satisfied. From the remaining equations it is clear that the

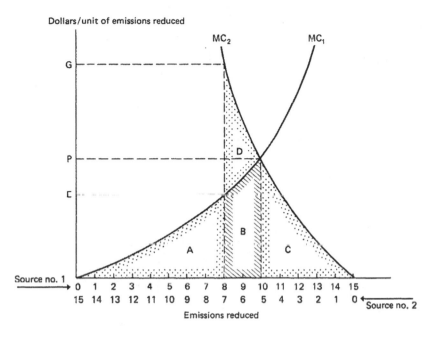

**FIGURE 2-1.** Cost-Effectiveness and the Emissions Permit System

permit system would yield the cost-effective allocation as long as $P = \lambda b$.

It turns out that the permit market would automatically yield this price. To see this, consider Figure 2-1, which can serve the twin purpose of showing how the price is determined as well as demonstrating in a graphical way why permit markets are cost-effective.

For simplicity, Figure 2-1 is drawn assuming only two sources, although the insights that we draw from it are valid for the larger set of J sources. For each source $\bar{e}_j - 15$, so the total uncontrolled emissions rate for the two firms is 30 units. The pollution target is assumed to be 15 units of allowed emissions, implying that the two sources together must reduce emissions by some 50% (15 units) if the target is to be met. The origin for the marginal cost of control for the first source ($MC_1$) in Figure 2-1 is the left-hand axis and the origin for the marginal cost of control for the second source ($MC_2$) is the right-hand axis. Notice that the desired 15-unit reduction is achieved for every point on this graph. Drawn in this manner, the diagram represents all possible allocations of the 15-unit reduction between the two sources. The left-hand axis, for example, represents an allocation of the entire control responsibility to the second source, while the right-hand axis represents a situation in which the first source bears the entire responsibility. All points in between represent different degrees of shared responsibility. It is easy to show that in a cost-effective allocation of the control responsibility between these two sources, the first source cleans up 10 units while the second source cleans up 5. The total variable control cost of this particular assignment of responsibility for the reduction is represented by the area A + B + C. Area A + B is the cost of control for the first source while area C is the cost of control for the second. Any other allocation would result in a higher total control cost.

Figure 2-1 also demonstrates the important proposition derived earlier. The costs of achieving a given reduction in emissions will be minimized if and only if the marginal costs of control are equalized across all emitters.[3] Because the marginal cost curves cross at the cost-effective allocation, they must be equal at that point. Given the presumed shape of the curves, marginal control costs are not equal at any other allocation.

What this analysis implies for emissions trading also can be seen in Figure 2-1. Suppose that prior to any emissions trading, the first source had to control 8 units. Since it has 15 units of uncontrolled emissions, this would mean it would be allowed 7 units of emissions. Similarly, if the second source were assigned 7 units of reduction, it would be allowed to emit 8 units.

Notice from Figure 2-1 that both firms have an incentive to trade since the marginal cost of control for the second source (G) is substantially higher than that for the first (E). The second source would lower its cost as long as it could buy emissions reduction credits from the first source at a price lower than G. The first source, meanwhile, would be better off as long as it could sell the credits for a price higher than E. Since G is greater than E, there are certainly grounds for trade.

Emissions trading would take place until the first source was controlling 10 units (2 more than originally), while the second source was controlling only 5 (2 less than originally). At this point, the permit price would equal P (since that is the value of a marginal unit of emissions reduction to both sources) and neither source would have any further incentive to trade. Not only would the permit market be in equilibrium, but the equilibrium would result in a cost-effective allocation of control responsibility.

As a result of the transfer of emissions permits equal to two units of emissions, the first source voluntarily controls more and the second source less. Allowing permits to be traded results in a lower cost of compliance without any change in total emissions. (In this example, area D represents the cost savings that result from allowing emissions trading.) For assimilative, uniformly mixed pollutants, emissions trading allows the control authority to achieve a cost-effective allocation despite its lack of knowledge about control costs. The regulatory authority merely would define the total level of emissions reduction, leaving the ultimate choice about control responsibility up to the sources. Meanwhile, sources have an incentive to adopt all cost-effective abatement opportunities.

## Non-Uniformly Mixed Assimilative Pollutants[4]

A second and somewhat more complex class of pollutants involves a relationship between emissions and the pollution target for which the location of the sources is crucial. A number of important air (and water) pollutants, such as total suspended particulates, sulfur dioxide, and nitrogen dioxide, fall within this classification.

For these pollutants, the policy target is specified in terms of a ceiling on the permissible ambient concentration of that pollutant measured at specific receptor locations.[5] Location is important because those concentrations are sensitive not only to the level of emissions but to the degree of source clustering as well. Clustered sources would be more likely to trigger violations of the standard than dispersed sources with the same aggregate emissions rate because with clustering the emissions would be concentrated in a smaller volume of air.

For this class of pollutants, the emissions–environmental quality relationship can be written as

$$A_i = \sum_{j=1}^{J} d_{ij}(\bar{e}_j - r_j) + a_i \qquad i = 1,...,I \qquad (2\text{-}13)$$

where $A_i$ is the concentration level measured at the $i$th receptor, $a_i$ is the background pollution level at that receptor, $d_{ij}$ is a transfer coefficient that translates emission increases or decreases by the $j^{th}$ source into changes in the concentration measured at the $i$th receptor, and $I$ is the number of receptors. The transfer coefficient expresses a steady-state relationship and takes into account such factors as average wind velocity and direction and the locations of sources and receptors as well as source stack heights.

The cost-effective allocation of a non-uniformly mixed assimilative pollutant is that allocation that minimizes the cost of pollution control subject to the constraint that the $I$-dimensional vector of predetermined concentration ceilings is met at all receptors.

Symbolically

$$\min_{r_j} \sum_{j=1}^{J} C_j(r_j) \qquad (2\text{-}14)$$

subject to
$$A_i \geq a + \sum_{j=1}^{J} d_{ij}(\bar{e}_j - r_j) \qquad i = 1,...,I$$

and
$$r_j \geq 0 \qquad j = 1,...,J$$

Once again, we can use the Kuhn-Tucker conditions to specify the nature of that allocation. For the problem defined in Equation 2-14, the Kuhn-Tucker conditions are

$$\frac{\partial C_j(r_j)}{\partial r_j} - \sum_{i=1}^{I} d_{ij}\lambda_i \geq 0 \qquad j=1,...,J \qquad (2\text{-}15)$$

$$r_j \left[ \frac{\partial C_j(r_j)}{\partial r_j} - \sum_{i=1}^{I} d_{ij}\lambda_i \right] = 0 \qquad j=1,...,J \qquad (2\text{-}16)$$

$$\bar{A}_i \geq a_i + \sum_{j=1}^{J} d_{ij}(e_j - r_j) \qquad i=1,...,I \qquad (2\text{-}17)$$

$$\lambda_i \left[ \bar{A}_i \geq a + \sum_{j=1}^{J} d_{ij}(e_j - r_j) \right] = 0 \qquad i=1,...,I \qquad (2\text{-}18)$$

$$r_j \geq 0; \quad i \geq 0 \qquad \begin{array}{l} j=1,...,J \\ i=1,...,I \end{array} \qquad (2\text{-}19)$$

Equation 2-15 states that for non-uniformly mixed assimilative pollutants in a cost-effective allocation, each source should equate its marginal cost of emissions reduction with a weighted average of the marginal cost of concentration reductions ($\lambda_i$) at each affected receptor. The weights are the transfer coefficients associated with each receptor. If the cost-effective pollutant concentration were lower than the ceiling at any receptor, equation 2-18 implies that the $\lambda_i$ associated with that receptor would be zero. This is referred to as a nonbinding receptor.

For any binding receptor, the associated $\lambda_i$ would be positive. Notice that for this class of pollutants, it is not the marginal costs of emissions reduction that are equalized across sources in a cost-effective allocation (as was the case for uniformly mixed assimilative pollutants); it is the marginal costs of concentration reduction at each receptor location that are equalized.

Seasonality in non-uniformly mixed assimilative pollutants not only affects the amount of allowable emissions (a characteristic it shares with uniformly mixed pollutants), but it also can affect transfer coefficients. Wind velocity and direction patterns frequently have a distinct seasonal component. Because of these seasonal effects, a cost-effective allocation can involve not only different seasonal levels of regional control but also different seasonal allocations of the emissions reduction responsibility among sources.

The permit system designed to yield a cost-effective allocation of the control responsibility for non-uniformly mixed assimilative pollutants is called an ambient permit system. Because this system involves a separate permit market associated with each receptor, each source would have to procure sufficient permits in each of the markets to legitimize its emissions rate.

Faced with the need to acquire permits from *I* markets, the source's decision on the level of control can be characterized as

$$\min_{r_j} C_j(r_j) = \sum_{i=1}^{I} P_i \left[ d_{ij}(\bar{e}_j - r_j) - q_{ij}^0 \right] \tag{2-20}$$

where $P_i$ is the price that prevails in the *i*th permit market and $q_{ij}^0$ is the *j*th source's initial allocation of concentration units at the *i*th receptor. The allocation that minimizes this cost is

$$\frac{\partial C(r_j)}{\partial r_j} - \sum_{i=1}^{I} P_i d_{ij} \geq 0 \qquad j-1,...,J \tag{2-21}$$

$$r_j \left[ \frac{\partial C(r_j)}{\partial r_j} - \sum_{i=1}^{I} P_i d_{ij} \right] = 0 \qquad j=1,...,J \tag{2-22}$$

$$r_j \geq 0 \qquad j=1,...,J \tag{2-23}$$

A comparison of equations 2-21 through 2-23 with equations 2-15 through 2-19 reveals $P_i = \lambda_i$ to be a sufficient condition for this permit system to produce the cost-effective allocation. As long as the control authority issues the appropriate number of permits for each receptor, the equivalence of supply and demand would ensure that $P_i = \lambda_i$ in each market.

Whereas emissions permits are defined in terms of allowable emissions rates, ambient permits are defined in terms of units of concentration at the receptor locations. The amount of allowed concentration at each receptor is determined by subtracting the background concentration $a_i$ from the concentration permitted by the ambient standard $\bar{A}_i$.

The denominations of these permits can be dictated by convenience. If the control authority wants a large number of permits to accommodate small sources, it could issue 1,000 permits, each denominated in units of 1/1,000 of the allowed concentration. If a smaller number of permits were desired, it could issue 100 permits, each worth 1/100 of the allowed concentration increase.

To explain how and why an appropriately designed emissions trading program would work in this context, it will simplify matters to consider initially a simple numerical example involving two sources and a single receptor (Table 2-1). Once the single-receptor case is understood, it is easy to deal with the added complexity of multiple receptors.

**TABLE 2-1.** Cost Effectiveness for Non-Uniformly Mixed Assimilative Pollutants: A Hypothetical Example

| Emissions units reduced | Marginal cost of emissions reduction (dollars per unit) | Concentration units reduced[a] | Marginal cost of concentration reduction[b] (dollars per unit) |
|---|---|---|---|
| | | Source 1 ($d_{11} = 1.0$) | |
| 1 | 1 | 1.0 | 1 |
| 2 | 2 | 2.0 | 2 |
| 3 | 3 | 3.0 | 3 |
| 4 | 4 | 4.0 | 4 |
| 5 | 5 | 5.0 | 5 |
| 6 | 6 | 6.0 | 6 |
| 7 | 7 | 7.0 | 7 |
| | | Source 2 ($d_{12} = 0.5$) | |
| 1 | 1 | 0.5 | 2 |
| 2 | 2 | 1.0 | 4 |
| 3 | 3 | 1.5 | 6 |
| 4 | 4 | 2.0 | 8 |
| 5 | 5 | 2.5 | 10 |
| 6 | 6 | 3.0 | 12 |
| 7 | 7 | 3.5 | 14 |

[a]Computed by multiplying the emissions reduction in Column 1 by the transfer coefficient.
[b]Computed by dividing the marginal cost of emissions reduction in Column 2 by the transfer coefficient.

Assume that the two sources have the same marginal cost curves for cleaning up the emissions as reflected by the identical first two corresponding columns of the table for each of the two sources. The main difference between the two sources is their location in relation to the receptor. Because the first source is assumed to be closer to the receptor, it has a larger transfer coefficient than the second (1 as opposed to 0.5).

Suppose the objective is to meet a given concentration target (3 units) at minimum cost and that with no control each source would emit 7 units of emission, resulting in 10.5 units of pollution concentration at the receptor. The relationship between emissions and concentration at the receptor for each source is given by column 3, while column 4 records the marginal cost of each unit of concentration reduced. The former is merely the emissions reduction times the transfer coefficient, while the latter is the marginal cost of the emissions reduction divided by the transfer coefficient (which translates the marginal cost of emissions reduction into a marginal cost of concentration reduction).

Suppose as part of a command-and-control approach to this pollution, the control authority required the first source to clean up 4 units of emission and the second source to clean up 7 units. This would meet the standard [(4)(1.0) + (7.0)(0.5) = 7.5] but would not be cost-effective.

The control costs have to be calculated to verify this. Costs can be calculated for each source control by adding up the marginal costs of emissions control associated with each control unit. The control cost for the first source would be $10^6$, while the control cost for the second source would be $28, producing a total cost for the two sources of $38.

If an emissions trading program were established, the sources would be free to trade concentration units. In particular, assume that the first source decided to clean up 6 units of emission rather than its command-and-control requirement of 4 units, selling the extra 2 units to the second source. By acquiring 2 concentration units from the first source, the second source could control 3 units of emission rather than 7. It would gain the right to emit two more concentration units, which translates into 4 more units of emission. The total concentration reduction (7.5) would be the same as that for the command-and-control policy, but the control cost of the emissions trading allocation would be lower. The emissions trading control costs for the first source would be $21, while for the second source it would be $6. The control costs for both sources taken together ($27) would be lower than those for the command-and-control policy ($38).

Allowing concentration trades reduces total control costs. The first source voluntarily undertakes more control because it can sell the concentration reduction credits for more than it costs to produce them. In this example, the credits would sell for $6 apiece since that is the price at which the marginal costs of concentration reduction are equalized for the two sources. At that price, the first source can sell for $12 what it costs $11 to produce. The second source would pay $12 for the two credits but would save $22 in control costs. The trade is in the interests of both parties. Furthermore, since the marginal cost of reducing the concentration to the desired level is equal for both sources, this trading equilibrium is cost-effective.

The extension of this system to the many-receptor case requires that a separate ambient permit be created for each receptor. The price prevailing in each of these markets would reflect the difficulty of meeting the ambient standard at that receptor. All things being equal, ambient permits associated with receptors in heavily congested areas could be expected to sustain higher prices than those affected by a relatively few emitters.

In multiple-receptor, ambient-permit markets, marginal emissions control costs would vary across sources. Sources with larger transfer coefficients would face higher marginal costs of emissions reduction.

This variation of costs with location would play a role in structuring cost-effective incentives by sending a signal to new sources that are deciding where to locate. The high control costs associated with heavily polluted areas would provide an incentive for heavy emitters to locate elsewhere. Even though pollution control expenditures are only part of the costs a firm considers when deciding where to locate, they are one factor.

The many-receptors case also can be used to think about the management of regional pollutants such as acid rain. Because these pollutants tend to be transported long distances by the prevailing winds, receptors far from the source can be affected by such emissions. The ambient permit system handles this geographic interdependency by requiring sources to purchase permits from all affected receptors—remote as well as local. All permits would be defined in terms of allowable concentration increases, with transfer coefficients used to relate emissions to concentration increases. The acid rain case is therefore conceptually no different than the general non-uniformly mixed case; only the proximity of the receptors is at issue. It is perhaps worth noting that the U.S. acid rain program does not use an ambient permit system. How the spatial aspects are handled by that program will be discussed in Chapter 4.

## Uniformly Mixed Accumulative Pollutants

The final class of pollutants under consideration contains substances that accumulate in the environment because their injection rate exceeds the assimilative capacity. In this section, it is assumed that location does not matter because for several important global pollutants that is the most reasonable specification, and because this assumption reduces the notational complexity. For some other accumulative pollutants (such as lead), location would matter.

For a uniformly mixed accumulative pollutant, a common form of the pollution target–emissions rate relationship can be described as

$$A_t = a_0 + \sum_{j=1}^{J} \sum_{k=1}^{t} (e_{jk} - r_{jk}) \qquad (2\text{-}24)$$

where the pollution level in any year $t$ is the simple addition of the initial pollution level ($a_0$) plus all emissions in the intervening years. This formulation implicitly assumes no assimilative capacity and a one-to-one correspondence between a 1-unit increase in emissions and a 1-unit increase in pollution; it also assumes no depreciation of the pollution stock over time.

Suppose that the policy objective was to ensure that $A$ would never be higher than some ceiling $\bar{A}$. Since the pollution–emissions relationship specified in equation 2-24 allows no means of reducing the size of the stock of accumulated pollutants, no further emissions are permitted once the ceiling is reached. The question of interest for this class of pollutants is not only how the cost-effective control responsibility is allocated among sources but how the degree of control changes over time as well.

The cost-effective allocation in this context is the one that has the lowest associated present value of control costs among all those allocations that satisfy the pollution constraint. The decision problem can be symbolically stated as

$$\min_{r_{jt}} \sum_{t=1}^{T} \sum_{j=1}^{J} \frac{C_j(r_{jt})}{(1+\rho)^{t-1}} \qquad (2\text{-}25)$$

subject to
$$\bar{A} \ge a_0 + \sum_{j=1}^{J}\sum_{t=1}^{T}(\bar{e_j} - r_{jt})$$

and
$$r_{jt} \ge 0 \qquad t = 1,...,T$$

where $\rho$ is the discount rate used to translate future costs into their present-value equivalents and $T$ is the length of the planning horizon. Relying once again on the Kuhn-Tucker conditions, it is possible to derive the conditions any cost-effective allocation will satisfy

$$\frac{\partial C_j(r_{jt})}{\partial r_{jt}}\frac{1}{(1+\rho)^{t-1}} - \lambda \ge 0 \qquad \begin{matrix} j=1,...,J \\[4pt] t=1,...,T \end{matrix} \qquad (2\text{-}26)$$

$$r_{jt}\left[\frac{\partial C_j(r_{jt})}{\partial r_{jt}}\frac{1}{(1+\rho)^{t-1}} - \lambda\right] = 0 \qquad \begin{matrix} j=1,...,J \\[4pt] t=1,...,T \end{matrix} \qquad (2\text{-}27)$$

$$\bar{A} - a - \sum_{j=1}^{J}\sum_{t=1}^{T}(\bar{e_j} - r_{jt}) \ge 0 \qquad (2\text{-}28)$$

$$\lambda\left[\bar{A} - a - \sum_{j=1}^{I}\sum_{t=1}^{T}(\bar{e_j} - r_{jt})\right] = 0 \qquad (2\text{-}29)$$

$$r_{jt} \ge 0; \ \lambda \ge 0 \qquad \begin{matrix} j=1,...,J \\[4pt] t=1,...,T \end{matrix} \qquad (2\text{-}30)$$

To the nonmathematical reader, this is probably the most intimidating set of equations yet, but as we found with the others, an intuitive understanding of what they convey is possible.

In those years when some, but less than complete, control is being exercised, the cost-effective control levels will satisfy

$$\frac{\partial C_j(r_{jt})}{\partial r_{jt}} = (1+\rho)\frac{\partial C_j(r_{jt-1})}{\partial r_{jt-1}} \qquad \begin{matrix} j=1,...,J \\[4pt] t=1,...,T \end{matrix} \qquad (2\text{-}31)$$

This implies that in a cost-effective allocation, marginal pollution control costs would rise over time at rate $\rho$ and the amount emitted would decline over time. In each time period, the marginal costs of control would be equalized across all sources.[7] Across time, the present value of marginal costs is equalized.

If $T$ is long enough, as $t$ increases eventually a year is reached in which the ambient constraint (equation 2-28) becomes binding and allowable emissions cease from then on.[8] The marginal control costs at that point are those associated with complete control; they are no longer necessarily equalized.

The permits yielding this cost-effective allocation do not have a time dimension; holders have complete freedom in timing their authorized emissions. In this market, the permits are an exhaustible resource; once used, they are withdrawn from circulation and the total number of allocated permits is limited.

Defining the appropriate number of permits is the first step that must be taken by the control authority to establish this market. The appropriate number is dictated by the environmental quality constraint; the amount of allowable emissions equals the pollution target minus the pollution already in the environment. Once these permits are issued, a market price would be established. The supply of unused permits diminishes over time (as some are used in the earlier years), while the demand for them increases. Prices rise to bring demand and supply into balance in each year.

Due to the demand and supply patterns, the rate of increase in permit prices would be equal to $\rho$, the rate of interest. This implies that sources respond to these rising prices by choosing that level of emissions control that equates marginal control cost and permit price. Since prices would rise at rate $\rho$, the rate of increase in marginal control costs also would be $\rho$, precisely what is required for the allocation to be cost-effective.

The notion that sources would conserve some permits for later years may not be obvious. What dictates conservation in this model? Any source that uses its permits too early forgoes the substantial value of those permits in the future as reflected in the higher prices. Myopic behavior raises the firm's costs unnecessarily, and that is inconsistent with our assumption that sources minimize the present value of their costs.

## The Role of Transactions Costs

The theory developed in the preceding sections is useful for two main purposes. First, it establishes that the conditions under which the incentives created by "appropriately designed" permit systems are compatible with cost-effectiveness. Second, it helps to establish what the appropriate design is.

This theory does not say that appropriately designed permit systems will achieve cost-effectiveness, only that under certain conditions the incentives created by the policy are compatible with that outcome. Whether or not a cost-effective outcome is achieved in practice depends on a number of conditions, such as the absence of either market power or lax enforcement, which will be discussed in succeeding chapters.

The ability of an emissions trading system to achieve full cost-effectiveness also depends upon how smoothly the market operates. One source of friction in these markets is transactions costs. Transactions costs are the costs, other than price, incurred in the process of exchanging goods and services. These include the costs of researching the market, finding buyers or sellers, negotiating and enforcing contracts for permit transfers, completing all the regulatory paperwork, and making and collecting payments.

Transactions costs are important because they can diminish the incentive to trade, and, once the decision to trade has been made, they can diminish the actual amount traded. They are also important because the magnitude of transactions costs can be affected by market design.[9]

Stavins (1995) published a careful theoretical study of the potential impact of transactions costs on permit markets. His study confirms the tremendous potential importance of transactions costs by demonstrating:

- The presence of transactions costs unambiguously decreases the volume of permit trading as long as the marginal cost functions are non-decreasing over the relevant ranges.

- Fixed transactions costs (those that are independent of the size of the trade) can affect whether or not a particular trade takes place but not its magnitude.

- Positive marginal transactions costs (those that are affected by the amount traded, such as a $10-per-ton-traded fee) reduce the amount exchanged in each trade and may diminish the number of trades.

- Whereas in the absence of transactions costs the final allocation of control responsibility will be cost-effective and independent of the initial allocation, in the presence of transactions costs the market equilibrium will depend on the initial allocation and in general will not be cost-effective.

Though these results temper to a considerable extent the optimistic conclusions presented in the previous theory, it is important to keep them in perspective. This theoretical work is extremely helpful in teasing out the implications of transactions costs, but how important these implications are in practice depends on their magnitude, a subject postponed until subsequent chapters. In addition, as Stavins (1995) points out, a trading system even with no trading likely will result in lower abatement costs than a command-and-control technology standard (since firms have more flexibility in choosing among the technological options).

These results suggest why comparing the theoretical properties of policies under similar assumptions and the implemented experience of policies under similar regulatory conditions can be a valid and useful exercise, but they also show why comparing the theoretical properties of one policy with the implementation experience of another would be highly misleading.

## The Role of Administrative Costs

Whereas transactions costs affect the buyers and sellers, a separate category of costs affects the regulators. Administrative costs, as defined here, are the costs of setting up and implementing a pollution control policy. As opposed to transactions costs, which are born by transactors, administrative costs are born by regulators.

Administrative costs for permit markets can be triggered by such activities as the need to retrain the staff, hold public hearings on proposals, establish the basis for the initial allocation, design the rules for trading, set up the registry system to keep track of permits, design a compliance system to match permit holdings to actual emissions, monitor emissions, and develop and implement strategies for managing noncompliance, among others. Both the nature and the intensity of the effort required to implement these activities are not the same for different policies.

Since administrative costs probably differ considerably across policy approaches, they should matter in policy choice. Unfortunately, they rarely are included in either theoretical or implementation experience comparisons.

Though the available implementation evidence on administrative costs is shared in the next chapter, for the purposes of this chapter, it is sufficient to lay out their effect in principle. Administrative costs can affect the efficiency of the policy choice and perhaps even its likelihood of implementation.

The efficiency of reform may be affected if the increase in administrative cost from moving to emissions trading from a command-and-control policy is larger than the savings in abatement costs. In this case, the cost savings from the greater flexibility experienced by emitters in how they meet the requirements is more than offset by the increased costs of setting up and running the program; total costs (the sum of abatement costs, transactions costs, and administrative costs) may rise when emissions trading is implemented. Models that examine only abatement costs ignore this important possibility.

Implementation feasibility may be affected if regulators play a crucial role in policy choice. Here, the sign of this effect is a bit more ambiguous. If regulators perceive that an emissions trading reform is likely to affect negatively the size of their staff or burden them with more work, they are likely to support the status quo. On the other hand, if the reform becomes the vehicle for a larger and more influential office, regulators could well support reform.

The point here is not to lay out the specific implications of administrative costs, but to emphasize that ignoring them could create a bias in the way policies are compared. Administrative costs are real and frequently not negligible.

# The Role of Technical Change

One expectation for emissions trading was that it would both encourage the development of new pollution control technologies and their adoption by regulated firms. Is it reasonable to believe that introducing transferability into the permit process would hasten the arrival and adoption of new technologies?

It is reasonable, but the incentives for technological adoption turn out to be a bit more complicated than might at first seem. How the permits are introduced into the market, for example, turns out to matter a lot. To get right to the conclusion of the theoretical work in this area, permits that are auctioned off provide both superior innovation and adoption incentives to command-and-control approaches, but the case is weaker for transferable permits that are distributed free.

The comparison of cost savings to firms under these policy approaches involves at least two considerations:

- First, adopting firms would accrue savings by using the cheaper technology to achieve the same level of control.
- Second, the adoption of this technology would lower the permit price (remember P = MC and MC is now lower). This fall in price conveys additional cost savings to buyers and fewer cost savings (than if the price had remained constant) to sellers.

The second effect is, of course, unique to emissions trading and its overall consequence is not obvious since buyers and sellers are affected in different directions.

Jung et al. (1996) specifically compare these possible outcomes to develop a ranking of the effects of instrument choice on the incentives for innovation and adoption. With emissions standards, the gain from adopting the new technology is simply the lower abatement cost from meeting the predetermined and fixed standard; no effects are transmitted to other regulated units. For emissions trading, however, the price effects on other parties must be considered.

Their analysis demonstrates that the sum of the abatement cost savings and gains and losses from the price effects and the resulting shift in control responsibilities among firms is greater in magnitude for an emissions trading system than for a set of command-and-control regulations where the permits are allocated free-of-charge; in other words, free distribution emissions trading provides greater incentive to adopt new technologies.

This same article goes on to demonstrate that the incentives for innovation are even greater if the permits are auctioned off rather than distributed free-of-charge. The logic behind this finding is that in an auction market, all firms are buyers (in contrast to a free-of-charge emissions trading system where some sources are buyers and others are sellers). Hence in an auction market, all emitters accrue the benefits from both the lower abatement costs and the lower permit prices.[10]

All results from theoretical models are, of course, contingent on the specification of the model. In this case, the model assumes that output prices (and levels) are unaffected by policy instruments in the permit market. How would the conclusions change if output effects were considered?

Fortunately, subsequent work by Montero (2002) and Bruneau (2004) sheds light on that question. The incentive of any firm in this setting to invest in innovative pollution control equipment is the function of two effects: the direct effect and the strategic effect.[11] The former captures returns to the firm that do not affect the output of competitors, while the latter captures returns that do have an affect on the output of competitors.

Strategic effects can arise in both the permit and output markets when either or both are imperfectly competitive but will be absent if both markets are perfectly competitive. In the permit market, when a firm invests in innovative pollution control equipment that causes a fall in the permit price, the lower price would affect rivals. In particular, the lower price would benefit rivals who are buyers of permits and disadvantage rivals who are sellers. In the output market, the lower permit price would lower costs of rivals and allow them to expand output. Anticipating these possible consequences allows firms to act strategically in making pollution control investment decisions. Strategic considerations can affect the incentive to innovate.

Formalizing these effects and considering differing degrees of imperfection in permit and product markets leads to the following results:

- With perfect competition in both permit and output markets, transferable permits (whether auctioned or distributed free-of-charge) provide more incentives for innovation than emissions standards when marginal production costs are increasing (Bruneau 2004, *1198*) and the same incentives when marginal production costs are constant (Montero 2002, *39*).

- With perfect (Cournot) competition[12] in the output market and imperfect competition in the permit market, with constant marginal production costs, auctioned permits, emissions standards, and emissions trading with free-of-charge distribution of permits all provide the same incentives for innovation (Montero 2002, *36*). With increasing marginal production costs, either form of permits provides more incentive to innovate than emissions standards (Bruneau 2004, *1198*).

- With imperfect (Cournot) competition in the output market and imperfect competition in the permit market, both emissions standards and auctioned permits provide more incentives to innovate than emissions trading with free-of charge permit distribution. The relative ranking of auctioned permits and emissions standards is ambiguous; it depends on the circumstances (Montero 2002, *31–32*).

- With imperfect (Cournot) competition in the output market and perfect competition in the permit market, emissions standards provide more innovation incentives than either auctioned permits or emissions trading with free-of-charge permit distribution (Montero 2002, *38–39*).

In general, considering output effects and the possibility of imperfect markets diminishes the otherwise apparently strong case that emissions trading in either form dominates emissions standards in terms of their ability to stimulate innovation. In principle, the degree to which this is true depends on marginal production costs. If marginal production costs are increasing, output effects become less important and the relative ranking of permit markets is enhanced (Bruneau 2004, *1192*). The next chapter considers evidence on how technical change has been influenced by emissions trading in practice.

## Summary

- The traditional command-and-control approach to air pollution control imposes a large information burden on control authorities. Because the magnitude of information required typically exceeds the amount of available information, control authorities have normally promulgated emissions standards that are not cost-effective.

- The specific design of cost-effective emissions trading systems depends crucially on the nature of the pollutant being regulated. Three common pollutant classes implying very different designs are: (1) uniformly mixed, assimilative pollutants; (2) non-uniformly mixed, assimilative pollutants; and (3) uniformly mixed, accumulative pollutants.

- A cost-effective allocation of uniformly mixed, assimilative pollutants could be achieved by an emissions permit system. For any geographic area, this system allows ton-for-ton trades among any sources in the airshed. The cost-effective emissions reduction credits would be defined in terms of an allowable number of emissions per unit time (such as 100 tons per year). Total cumulative emissions would not be controlled.

- For non-uniformly mixed assimilative pollutants, an ambient permit system would yield the cost-effective allocation of control responsibility. For each airshed, this system requires considering the spatial effects of trades on each receptor site. The cost-effective emissions reduction credits in this approach would be defined in terms of allowable concentration increases at specific receptor locations. Separate permits, which could be banked or sold independently, would be associated with each receptor. Total cumulative emissions would not be controlled with this system either.

- Uniformly mixed accumulative pollutants can be cost-effectively controlled using a cumulative emissions permit system. Firms would be free to bank permits and they would be traded on a one-for-one basis. The number of permits would limit total cumulative emissions.

- In the presence of transactions costs, the relative cost-effectiveness of emissions trading over emissions standards is diminished but not eliminated unless the transactions costs are so high as to preclude any trading.

- The cost-effectiveness of emissions trading also may be affected if the administrative cost increases as a result of moving to emissions trading from a command-and-control policy. In this case, the cost savings from the greater flexibility experienced by emitters in how they meet the requirements potentially could be offset by the increased costs of setting up and running the emissions trading program. If the administrative cost differential is positive and large enough to offset the abatement cost savings, total costs (the sum of abatement costs, transactions costs, and administrative costs) could even rise when emissions trading is implemented.

- The regulatory information burden necessary to achieve a cost-effective allocation of the control responsibility is smaller with emissions trading systems than with the command-and-control approach. Of the three considered permit approaches, ambient permit systems impose a higher burden than the other two permit systems since they require the use of transfer coefficients that govern the allowable transfers among sources.

- When both permit and output markets are perfectly competitive, permit markets in principle produce stronger incentives for innovation than emissions standards. Generally, auctioning permits off provides more innovation incentives than gratis distribution.

- In the presence of imperfect permit and output markets, when strategic considerations affect the pollution control investment behavior of firms, the dominance of permit markets in providing incentives for innovation is diminished and in some specific cases even reversed.

## Notes

1. In subsequent chapters that review the state of the art for various aspects of permit design, studies that use an efficiency, rather than a cost-effectiveness, framework also are included when that framework reveals useful insights about the operation of emissions trading markets. Since cost-effectiveness has been the main guiding force for design to date, it is the focus here for both historical and practical reasons.

2. The theory for this case was first developed by Baumol and Oates (1971).

3. This is strictly true only if all sources are assigned some control responsibility in a cost-effective allocation. Two possible exceptions occur when no control is needed ($\lambda = 0$) or when a source (say the $j_{th}$) is so costly to control that it is assigned no control responsibility ($r_j = 0$).

4. The theory for this case was first developed by Montgomery (1972).

5. U.S. law specifies that these ceilings are to be met everywhere and not merely at specified monitored locations. As a practical matter, however, studies have shown that a relatively few monitors effectively can cover a particular region of interest (Ludwig, Javitz, and Valdes 1983).

6. The marginal costs of the first, second, and third unit of reduction for the first source are, respectively, $1, $2, and $3, for a 3-unit control cost of $6. With a marginal cost of $4, the 4-unit control cost is $10. The calculations for the second source are similar.

7. This analysis presumes smooth cost functions. In practice, it may be necessary to purchase emissions control in large discrete increments. In this case, the source would make a few large investments, producing more control in the earlier years than expected on the basis of smooth cost functions. Eventually, even in this case, total control would be necessary; only the path by which this outcome would be reached would differ.

8. Emissions cease in linear demand models. With nonlinear demand, the emissions will asymptotically approach zero.

9. One simple example of this point is illustrated by the decision in many markets to ensure that transactions prices are transparent to potential transactors. Having a clear sense of the market price reduces uncertainty considerably for both buyers and sellers and makes planning abatement investments much easier. As the subsequent discussion of market design makes clear, some markets have incorporated this insight while others have not.

10. Jung et al. (1996) also demonstrate that auctioned permits provide more incentive for innovation than either an emissions tax or subsidy.

11. This characterization was first proposed in Tirole (1988).

12. In Cournot, perfect competition firms take the other firms' decisions into account in making their innovation investment decision, but they do not affect output price. In Cournot, imperfect competition firms can affect output price.

# 3

# The Consequences of Emissions Trading

The theory in the preceding chapter provides a strong a priori case for believing that the traditional command-and-control approach to emissions control is not (and cannot become) cost-effective. The amount of information required by the regulatory authority to establish a set of cost-effective emissions standards is so high as to preclude a cost-effective outcome. In contrast, cost-effective permit systems are, in theory at least, a distinct possibility because the information requirements for initiating an emissions trading system are lower.

As interesting as these theoretical results are, they provide an incomplete guide to regulatory reform. Because any change in policy has its own set of costs, it is difficult to overcome the inertia of the status quo. New grounds for legal challenge are exposed. Bureaucratic staffs trained in one set of procedures must learn new ones. The comfort of familiarity is lost to both regulators and sources.

In order to overcome this inertia, successful reforms usually involve either a small departure from the traditional approach to hold down the costs of change as perceived by the participants, or staffs must be able to quickly come up to speed by drawing upon the experience of others who have implemented similar programs. The earliest EPA emissions trading program, established before other programs could supply information, satisfied the "small departure" condition because it built upon the traditional approach. Emissions reduction credits were a complement to, not a replacement for, the traditional approach.

Reforms that are successful in overcoming the inertia of the status quo must also promise benefits substantial enough to outweigh any frictional costs of moving away from a more traditional approach. The theoretical models presented in Chapter 2 made it clear that the benefits to be realized from this reform (in the form of reduced compliance costs) are positive, but these theoretical models are not necessarily sufficient to shed light on the important question of whether or not the benefits are substantial. Answering that question requires an appeal to a different kind of evidence.

This chapter surveys two kinds of evidence that respond to that need: (1) ex ante computer simulations comparing least-cost to command-and-control allocations; and (2) ex post evaluations that are based upon actual implementation experience. The objective is not only to discover whether the perceived ex ante potential benefits were substantial enough to motivate change but also discover whether the expectations created by these studies were born out by the ex post evaluations. The reliability of these estimates as well as their magnitudes will be explicitly considered.

## The Nature of the Evidence

### Ex Ante Simulation Studies

One way to gain information on the cost inefficiency of the traditional approach is to compare the command-and-control allocation of control responsibility with a least-cost allocation using a computer simulation model. The models used to perform these simulations depend heavily on the theoretical work covered in Chapter 2. For each type of pollutant, the optimality conditions are used as the basis for algorithms designed to find the least-cost allocation of control responsibility for the specific pollutant and geographic area being investigated.[1]

The chief virtue of these simulation models is that they allow the analyst to examine counterfactual situations. As will be clear in subsequent sections of this book, a number of constraints can operate on emissions trading programs. These arise from the statutes, from court decisions, or simply from the implementation rules that flow from the bureaucracy. Whatever the source, it is useful to know how seriously they jeopardize the degree to which the reforms can achieve the objectives established for them.

Interestingly, ex ante analysis is uniquely helpful here. Ex post experience with emissions trading is not much help in pinning down the cost of various restrictions since those restrictions usually apply to every transaction. Discovering the unique effects of any particular constraint requires a comparison of the allocations with and without the constraint. In the absence of controlled experiments, this comparison is possible only with simulation models.

The estimates of potential cost savings from ex ante simulation models are enormously useful when the simulations provide the basis for deciding whether to allow emissions trading and, if so, what forms the program should take. For this purpose, what is important is the order of magnitude of the relative cost savings and how sensitive this order of magnitude is to various design changes, not the absolute values of the cost associated with any particular scenario. As we shall see, for the most part the magnitudes have been so large as to sustain the desirability of reform, even allowing for measurement errors.

## Limits of Ex Ante Studies

The ability to obtain answers to "What if . . . ?" questions is a substantial virtue, but ex ante simulation models do have their drawbacks. Perhaps the chief drawback is that they typically deal with an idealized situation. Rather than simulating the actual workings of permit markets, these simulations find the least-cost allocation for meeting a particular standard. Equating this particular allocation with a permit market equilibrium is, in principle, valid for perfectly functioning markets; it is not necessarily valid for less-than-perfect markets.

Actual cost savings from a reform are likely to differ from the ex ante estimates of potential savings for three main reasons: (1) the cost of the least-cost allocation may be measured with some error; (2) the cost of the command-and-control allocation may be measured with some error; and (3) the implemented emissions trading program can differ considerably from the idealized programs modeled by these analyses. Even simulation models that perfectly capture abatement costs can only tell us how much cost-savings is possible (an upper bound), not how much actually would be achieved by emissions trading.

One specific source of concern relates to the starting point for the analysis. Some simulations comparing the command-and-control allocation with the least-cost allocation implicitly assume a "no control" benchmark. Both the command-and-control benchmark and emissions trading are examined as if each were the first policy implemented. While this is an appropriate comparison in its own right, it is not appropriate to interpret the resulting cost difference as a measure of the potential cost savings that could be achieved by moving from an existing command-and-control system to an emissions trading program.

The distinction is important. When command-and-control regulations are already in place prior to the introduction of emissions trading, sources would have already purchased and installed a good deal of durable capital equipment to comply with those regulations. Because this allocation of capital is fixed, in the short term it constrains the feasible set of reallocations of control responsibility. Savings that could have been achieved by a reform package imposed on firms making their initial investments in abatement equipment are not, in general, an adequate estimate of the savings that would be achieved when firms have to replace their existing capital.

The computer estimates would represent a correct calculation of the maximum cost savings from the reform package if and only if all the capital equipment in the command and control strategy could be, at no cost, disassembled at those sources needing less control in the least-cost solution and reassembled at those sources needing more. Since as a practical matter this cannot be done, the existing configuration of control equipment is a constraint on emissions trading in the short run that the models fail to recognize. Because in many programs emissions trading does not face a clean slate in assigning control responsibility, the maximum cost savings actually achieved could be smaller, perhaps considerably smaller, than predicted by the models.

The degree of overstatement would depend on how much capital equipment for control is already in place when the reforms are initiated in a particular location and how much of that has to be replaced. The more inappropriate capital equipment that is already in place, the larger the bias will be.

This analysis suggests that in the short run, ex ante estimates could overestimate the cost savings. Interestingly, in the long run—a period of time sufficiently long enough that the menu of control possibilities can change and sources can react to the expanded set of possibilities—the cost savings calculated by these static models could underestimate the maximum actual savings.

Static models provide a snapshot of the situation at a moment in time, while dynamic models follow the evolution of the system over time. Because of their timelessness, static models are incapable of capturing the role of emissions trading in influencing the evolution of the economic system (the stimulating role) and in responding to that evolution (the facilitating role). Both are enduring roles for emissions trading after the initial flurry of trading activity has passed.[2]

In addition to its role in affecting innovation (covered in Chapter 2), emissions trading also encourages the earlier shutdown of plants that, for whatever reason, are likely to close sooner or later anyway. When closed plants can sell their permits (a possibility in some, but not all, programs), this produces an extra source of revenue compared to a command-and-control system.

An interesting property of this revenue is that it would be higher in markets where the demand for permits from expanded or new sources is higher. Therefore, emissions trading would provide the most encouragement for shutdowns in those areas where other facilities are waiting to take up the slack. If the labor intensity of the new and old sources were the same, the impact of the shutdown on employment would be small because it would be offset by the gains in employment associated with the new source. Thus, emissions trading would provide the most shutdown encouragement when the detrimental effects on employment were small.

## Ex Post Analyses

A second source of information for analysis is provided by a retrospective look at the data after a program has been functioning for a sufficient period of time. This provides an opportunity to get a look at actually functioning markets and to get a handle on such as aspects as the magnitude of administrative and transactions costs, as well as abatement costs.

In principle, establishing how well a program has worked in actual application seems a simple matter. In practice, it is more complicated than it seems. As a result, reasonable people viewing the same experience can come to different conclusions. Therefore, before delving into the evidence provided from ex post evaluations, it is reasonable to take a close look at the ex post evaluation process itself.

Ex post studies typically rely upon some or all of three different economic criteria: efficiency, cost-effectiveness, or market effectiveness. Relying on different concepts creates the potential for drawing different, even contradictory, conclusions.

Efficiency, or its typical operational formulation, maximizing net benefits, examines whether or not the policy maximizes net benefits. Naturally, this requires a comparison of the costs of the program with all the benefits achieved, including the value of reduced pollution (Burtraw et al. 1998). Conducting this kind of evaluation is time and information intensive and in practice is quite rare.[3] An alternative form, which is somewhat less rare, is simply to compare the present value of net benefits for the program with the net benefits from some predefined alternative.

A different and more common evaluation approach relies on cost-effectiveness (Farrell et al. 1999). As noted in the second chapter, this approach typically takes a predefined environmental target as a given (such as an emissions cap or ambient standard) and examines whether the implemented program minimizes the cost of reaching that target. Another version involves comparing the cost of reaching the target with emissions trading to the cost of reaching the target with the next most likely alternative. This approach, of course, compares the program not to an optimal benchmark but to a pragmatic (most likely) benchmark.

Finally, a number of evaluations focus on whether the market architecture is effective or not in terms of facilitating trading. In the absence of an initial allocation that happens to mimic the cost-effective allocation, transactions costs (see Chapter 2) and market power (see Chapter 7) can inhibit trade and prevent a market from achieving the target at minimum cost. A number of studies (Ellerman 2003; Harrison 2004) examine market effectiveness.[4] These studies use both qualitative and quantitative assessments to gauge not only the outcomes of the process but also the effectiveness of the process itself.

## Defining the Appropriate Benchmark

Since many ex post evaluations compare the environmental policy to an alternative pragmatic benchmark, defining the appropriate benchmark is crucial. It also is difficult.

Emissions trading, of course, usually is not implemented in a vacuum. It frequently complements, or is intertwined with, other policies. For example, the U.S. Sulfur Allowance Program operates within the more general framework of sulfur oxide regulation established by the National Ambient Air Quality Standards. The RECLAIM program in California was affected dramatically by the problems that flowed from the electric deregulation program. The interdependence of these programs makes it difficult to disentangle the unique effects of an emissions trading policy and to draw implications for how the policy might work in a different policy environment.

Ellerman (2003), for example, points out that the baseline for the evaluation of the Sulfur Allowance Program conducted at MIT is based on the assumptions that: (1) the emissions rate observed at affected units would not have changed from the pre-emissions trading rates; and (2) the heat input or demand observed at affected units in each year is not affected by the emissions trading program. These assumptions have the effect of making the estimated counterfactual emissions sensitive to changes in demand at individual units and in the aggregate. However, this approach will err to the extent that other environmental regulations or changes in relative fuel prices changed emissions rates from the observed pre-trading rate or that emissions trading changed the dispatch of affected units either individually or in the aggregate.

Developing counterfactuals about costs is necessarily more subjective than developing counterfactuals about emissions since relative costs depend directly on the degree of inefficiency assumed in the hypothetical alternative regime. Since that regulatory regime doesn't exist, it is not always easy to figure out what it might have been.

Finally, the easiest counterfactual, constructed by simply extrapolating past trends, might prove misleading. The future is rarely a mirror of the past. Important aspects of the problem, such as control technologies, economic conditions, or scientific knowledge, can change and the system naturally would respond to those changes even in the absence of an emissions trading program.

A different approach is taken in Harrington et al. (2004). They compare the actual outcomes from controlling similar pollutants in the United States and in Europe using different policy instruments. The fact that the United States and Europe have made different policy choices allows the two implementation experiences to be compared with each other, rather than with a hypothetical baseline.

The difficulty with this approach is that not all of the identified differences can be attributed purely to the instruments. Europe and the United States have different regulatory cultures and even more different economic environments. Part of the differences in outcomes inevitably must reflect those differences rather than the instrument choice.

## The Scope of the Evaluation

It is important to note that evaluation difficulties arise not only from specifying what policies to include in (or omit from) the counterfactual benchmark but also in isolating the degree to which changes in outcomes are endogenous or exogenous to the policy change. To the extent that the introduction of the program influences outcomes that normally are considered outside of the scope of the analysis, important aspects may be missed. And, as elaborated below, several circumstances can be identified where the apparent effects of the program transcend normal evaluation boundaries.

Clearly the outcome of the Sulfur Allowance Program, for example, has been heavily influenced by dramatic changes in scrubber technology and by the markedly enhanced rail availability of low-sulfur coal from the western United States. Would these events have occurred in the absence of the Sulfur Allowance Program (and therefore should be in the counterfactual) or were they the result of the program (and therefore should not be in the counterfactual)? Definitive conclusions about the effectiveness of this program depend on the answers to those questions; yet achieving consensus about the answers remains elusive.

As described below in the section dealing with substantive ex post results, environmental outcomes that cost-effectiveness evaluations normally may assume to be the same under alternative policy regimes may not be the same at all. One example of such variability involves the degree and cost of monitoring and enforcement.

Do monitoring and enforcement costs rise under emissions trading programs? Although a detailed consideration of this question is postponed until Chapter 8, the answer depends both on the level of required enforcement activity (greater levels of enforcement obviously cost more) and on the degree to which existing enforcement resources are used more or less efficiently. Even higher enforcement costs may not, by themselves, be definitive if they turn out to be more easily financed from the lower costs resulting from emissions trading.[5]

Evaluations of tradable permit programs must have a sufficiently large scope to take "external" effects into account. Resources controlled by the permit program frequently are not the only resources affected.

In air pollution control, several effects transcend the normal boundaries of the program. The climate change program provides the most dramatic example of this. As is now widely recognized (Hartridge 2003; Ekins 1996), the control of greenhouse gases will result in substantial reductions of other pollutants as a side effect. Ex post studies that ignore this additional reduction produce a biased evaluation.

Other more detrimental effects include the clustering of emissions in space or time. For some but not all pollutants, the location of the emission can matter (Tietenberg 1995). Any cost-effectiveness analysis that does not account for the spatial or temporal heterogeneity in emissions may be defining effectiveness incorrectly.

Finally, the scope of the evaluation needs to take into consideration both the interaction with policy instruments outside the permit market as well as the consequences of those interactions that also fall outside the permit market. Tax interaction effects represent the classic example of this point. As elaborated in more detail below, when permit markets are imposed on an economy with distortionary taxes, the permits can intensify the distortions— effects that would not be apparent in an analysis focusing solely on the permit market.

## The Timing of the Evaluation

Evaluations can be conducted at any time during the life of the program, but when they are conducted will affect what they find and how the findings may be most usefully interpreted. Most of these programs evolve considerably over their lifetimes. As will be demonstrated subsequently, not only do participants and administrators experience a considerable amount of "learning by doing" as the program matures, but design parameters frequently are altered in light of early experience. Recognizing this potential for evolution implies that early evaluations may not provide much insight about the ultimate success or failure of the program since dramatic change is common.

Consider a specific example. In the Sulfur Allowance Program, firms looked first to internal trades and to process adjustments rather than to external trades to meet requirements. Thus, in the early stages of this program the equalization of marginal costs, the focus of the theory, proved to be less important than the opportunity to exploit technological flexibility within the firm (Bohi and Burtraw 1997). As the program matured, however, the tendency for external trades increased.

## The Role of Administrative Costs

Although a complete ex post evaluation would examine how policy choice affects administrative costs as well as abatement cost, most do not. Although most published case studies don't shed much light on administrative costs, some case studies do. As demonstrated below, they demonstrate that both the number and nature of public administration tasks can change with the adoption of a tradable permits approach. Therefore, ex post studies that fail to consider administrative costs omit by design a potentially important comparative element.

# Ex Ante Evaluations: The Evidence

From the theory in Chapter 2, we know that, in principle, appropriately designed emissions trading systems are capable of achieving a cost-effective allocation. In practice, the savings registered by emissions trading programs will be determined not only by the potential cost savings, which is measured as the deviation of the cost of the command-and control allocation from the lowest possible cost of achieving the same pollution target, but also by the degree to which the costs resulting from actual emissions trading programs approximate the least-cost solution.

The magnitude of potential cost savings depends on many local circumstances, such as prevailing meteorology, the locational configurations of sources, stack heights, program design characteristics, and how costs vary with the amount controlled. Over the last several decades, several simulation models have been constructed that integrate these factors for specific pollutants in specific airsheds. A representative group of these studies is presented in Table 3-1.

Since for a variety of reasons the estimated costs cannot be directly compared across studies, the potential cost savings are presented as the percentage reduction in costs that could be achieved by the least-cost allocation of responsibility compared to the command-and-control benchmark. To ensure comparability for each study, both the command-and-control allocation and the least-cost allocation are defined in terms of the same set of ambient standards (but not necessarily the same emissions loadings).

The vast majority of these studies find very large possible savings in abatement costs. The Seskin, Anderson, and Reid (1983) and Spofford (1984) estimates are sufficiently larger than the others as to deserve special mention. The former study is concerned with meeting a stringent, short-term standard. The concentrations at the most polluted receptors are heavily influenced by the location of the emissions. Because the least-cost strategy takes emission location into account while the command-and-control strategy does not, it is able to meet the ambient standard at a significantly lower cost.

The Spofford study finds a high potential cost savings for particulate control because of the manner in which it treats area sources in the command-and-control allocation. Application of a uniform percentage reduction to all area sources, as well as to point sources, turns out to be a very expensive way to control pollution. Some 89% of the total regional control costs associated with the command-and-control allocation are accounted for by area sources, while they account for only about 12% of total emissions.[6]

## The Causes and Consequences of Low Potential Cost Savings

Under what conditions might the potential cost savings be small? One clue can be found in the Los Angeles study, which estimated that the potential cost savings from using emissions trading in that city would be very small. A closer look at that study begins to suggest a circumstance in which the normally powerful cost-reducing properties of emissions trading may be less effective.

The authors suggest several reasons why the estimated cost savings were so low in Los Angeles, where a great deal of sulfate pollution is occurring.[7] First, in contrast to other areas, the command-and-control strategy in California did not at that time include scrubbers, a very expensive approach. Had California required scrubbers, the potential cost savings would have been higher.

Another reason of more general applicability is that the amount of control required is so great that every source is forced to control as much as is economically feasible. By definition, little further control can be undertaken. Since further reductions are necessary for emissions trading to take place, the immediate opportunities for cost-saving transfers (the ones measured by static simulations) are extremely limited. Therefore, as long as the control authority has to impose emissions standards that are close to the limit of technological feasibility, in the absence of any changes in circumstances the divergence between the command-and-control and least-cost allocation would be small.

While the divergence has to be zero at the maximum control point (assuming the command-and-control allocation does not err by imposing infeasible reductions), it is not clear that the potential cost savings would decline monotonically with the degree of control required. Atkinson and Lewis (1974, 245), for example, show that particulate control in St. Louis exhibits cost savings over a considerable range of increasingly stringent ambient standards.

Other studies generally have presented less complete information on the full range of possible ambient standards, concentrating instead on the more stringent end of the range. They generally find that in this range, increasing stringency implies reduced potential cost savings. Spofford (1984, 57, 66, and 77) finds this to be the case for both particulates and sulfur dioxide, while Maloney and Yandle (1984, Table V) find it to be true for hydrocarbon control.

## Interaction Effects

Most of the studies discussed so far deal with partial equilibrium effects in the sense that they consider permit markets in isolation from other markets. Yet transactions in the permit market do have effects on other markets—input and output markets, for example. Interestingly, when these effects are considered in the context of general equilibrium models, the results suggest not only that these effects can be quantitatively important but even that they affect conclusions about the desirability of emissions trading as a policy tool.

In particular, these studies find that emissions permit systems can drive up the output price from polluting industries, a product-market effect that can exacerbate other economic distortions derived from the tax system. Intensifying these distortions increases the costs of emissions trading policies compared to the costs that would have been estimated using a partial equilibrium analysis.

Central-case estimates suggest that the net effect of such interactions is to increase the costs of emissions trading slightly when all of the permits are auctioned and by much more if they are not (e.g., Burtraw et al. 1998; Parry et al. 1999, or Goulder et al. 1999). These effects can in some cases cause emissions

**TABLE 3-1.** Empirical Studies of Air Pollution Control

| Study and year | Pollutants covered | Geographic area | CAC benchmark | Assumed pollutant type | % Cost saving |
|---|---|---|---|---|---|
| Atkinson and Lewis (1974) | Particulates | St. Louis metro area | SIP regulation | Non-uniformly mixed assimilative | 83.4[a] |
| Hahn and Noll (1982) | Sulfates | Los Angeles | California emissions standards | Non-uniformly mixed assimilative | 6.6 |
| Krupnick (1983) | Nitrogen dioxide | Baltimore | Proposed RACT regulation | Non-uniformly mixed assimilative | 83.3[b] |
| Seskin, Anderson, and Reid (1983) | Nitrogen dioxide | Chicago | Proposed RACT regulation | Non-uniformly mixed assimilative | 93.1 |
| McGartland (1984) | Particulates | Baltimore | SIP regulations | Non-uniformly mixed assimilative | 76.1 |
| Spofford (1984) | Sulfur dioxide | Lower Delaware Valley | Uniform percentage reduction | Non-uniformly mixed assimilative | 43.9 |
| | Particulates | Lower Delaware Valley | Uniform percentage reduction | Non-uniformly mixed assimilative | 95.5 |
| Harrison (1983) | Airport noise | United States | Mandatory retrofit | Uniformly mixed assimilative | 41.9[c] |
| Maloney and Yandle (1984) | Hydrocarbons | All domestic DuPont plants | Uniform percentage reduction | Uniformly mixed assimilative | 75.9[d] |
| Palmer, Mooz, Quinn, and Wolf (1980) | Chlorofluoro-carbon emissions from nonaerosol applications | United States | Proposed emission standards | Uniformly mixed assimilative | 49.0 |
| Krupnick (1986) | Nitrogen dioxide | Baltimore | RACT/Least-Cost | Non-uniformly mixed assimilative | 24.3[e] 69.1[f] 95.7[g] |
| Farrell, Carter, and Raufer (1999) | Nitrogen oxides | Northeastern United States | Initial allocation of the $NO_x$ Budget | Non-uniformly mixed assimilative | 46.6 |
| O'Ryan (1996) | Particulates | Santiago, Chile | Uniform concentration standard | Non-uniformly mixed assimilative | 83.4[h] 41.2[i] |

| Johnson and Pekelney (1996) | Southern California | Sulfur and nitrogen oxides | Air Quality Management Plan | Non-uniformly mixed assimilative | 57.3 |
| Førsund and Naevdal (1998) | Europe | Sulfur dioxide | Second Sulphur Protocol | Non-uniformly mixed assimilative | 57.3 |

*Definitions*: CAC = Command and control, the traditional regulatory approach; SIP = State implementation plan; RACT = Reasonably available control technologies, a set of standards imposed on existing sources in nonattainment areas.

*Sources*: Scott E. Atkinson and Donald H. Lewis, "A Cost-Effectiveness Analysis of Alternative Air Quality Control Strategies," *Journal of Environmental Economics and Management* vol. 1, no. 3 (November 1974) p. 247; Robert W. Hahn and Roger G. Noll, "Designing a Market for Tradeable Emission Permits," in Wesley A. Magat, ed., *Reform of Environmental Regulation* (Cambridge, Mass., Ballinger, 1982), tables 7-5 and 7-6, pp. 132-133; Alan J. Krupnick, "Costs of Alternative Policies for the Control of NO₂ in the Baltimore Region" (unpublished Resources for the Future working paper 1983) table 4, p. 22; Eugene P. Seskin, Robert J. Anderson, Jr., and Robert O. Reid, "An Empirical Analysis of Economic Strategies for Controlling Air Pollution," *Journal of Environmental Economics and Management* vol. 10, no. 2 (June 1983) tables 1 and 2, pp. 117 and 120; Albert Mark McGartland, "Marketable Permit Systems for Air Pollution Control: An Empirical Study," (Ph.D. dissertation, University of Maryland, 1984) table 4.2, p. 67a; Walter O. Spofford, Jr., "Efficiency Properties of Alternative Source Control Policies for Meeting Ambient Air Quality Standards: An Empirical Application to the Lower Delaware Valley" (unpublished Resources for the Future discussion paper D-118, February 1984) table 13, p. 77; David Harrison, Jr., "Case Study 1: The Regulation of Aircraft Noise," in Thomas C. Schelling, ed., *Incentives for Environmental Protection* (Cambridge, Mass., MIT Press, 1983) tables 3.6 and 3.16, pp. 81 and 96; Michael T. Maloney and Bruce Yandle, "Estimation of the Cost of Air Pollution Control Regulation," *Journal of Environmental Economics and Management* (1984, forthcoming) table IV; Adele, R. Palmer, William E. Mooz, Timothy H. Quinn, and Kathleen A. Wolf, *Economic Implications of Regulating Chlorofluorocarbon Emissions from Nonaerosol Applications*, Report #R-2524-EPA prepared for the U.S. Environmental Protection Agency by the Rand Corporation (June 1980) table 4.7, p. 225.Alan J, Krupnick, "Costs of Alternative Policies for the Control of NO₂ in the Baltimore Region." *Journal of Environmental Economics and Management* vol 13, no 2 (June 1986) tables II and III, pp. 193-194; Farrell, A., R. Carter, and R. Raufer, "The NOₓ Budget: Market-Based Control of Tropospheric Ozone in the Northeastern United States" *Resource and Energy Economics* vol 21, no2, p. 103-124; O'Ryan, R. "Cost-Effective Policies to improve Urban Air Quality in Santiago, Chile" *Journal of Environmental Economics and Management* vol 31, no 3 (November 1996) p. 308; Scott L. Johnson and David M Pekelney "Economic Assessment of the Regional Clean Air Incentives Market: A New Emissions Trading Program for Los Angeles" *Land Economics*, vol 72, no 3 (August, 1996): 277-297; Førsund, F. R. and E. Naevdal. "Efficiency Gains Under Exchange-Rate Emission Trading." *Environmental and Resource Economics* vol 12, no 4 (1998) p.417.

a. Based on a 40 g/m³ at worst receptor.

b. Based on a short-term, 1-hour average of 250 g/m³.

c. Because it is a benefit-cost study instead of a cost-effectiveness study, the Harrison comparison of the command-and-control approach with the least-cost allocation involves different benefit levels. Specifically, the benefit levels associated with the least-cost allocation are only 82 percent of those associated with the command-and-control allocation. To produce cost estimates on more comparable benefits, as a first approximation the least-cost allocation was divided by 0.82 and the resulting number was compared with the command-and-control cost.

d. Based on 85 percent reduction of emissions from all sources.

e. Achieves air quality of 250 µg/m³.

f. Achieves air quality of 375 µg/m³.

g. Achieves air quality of 500 µg/m³.

h. Based on a 60% reduction in concentrations.

i. Based on a 90% reduction in concentrations.

trading to be less efficient than command and control (Goulder et al. 1999) or cause narrowly defined emissions trading programs (that leave some sectors completely unregulated) to be more efficient than broader programs (Parry and Williams 1999).

# Ex Post Evaluations: The Evidence

This assessment of the outcomes of these systems focuses on three major categories of effects. The first is implementation feasibility. A proposed policy regime cannot perform its function if it cannot be implemented or if its main protective mechanisms are so weakened by the implementation process that it is rendered ineffective. What matters to policymakers is not how a policy regime works in principle but how it works in practice. The second category seeks to answer the question, "How much environmental protection did it offer not only from the targeted pollutant, but to the environment in general?" Finally, this type of survey examines the economic effects of emissions trading programs.

## *Implementation Feasibility*

Until recently, the historic record seemed to indicate that resorting to emissions trading usually occurred only after other, more familiar approaches had been tried and failed. In one view of this evolution, the adjustment costs of implementing a new system with which policy administrators had little personal experience were apparently perceived as so large that they could only be justified when the benefits had risen sufficiently to justify the transition (Libecap 1990).

This review finds some support for that view, particularly in the earlier years. The offset policy, introduced in the United States during the 1970s, owes its birth to an inability to find any other policy to reconcile the desire to allow economic growth with the desire to improve the quality of the air (Tietenberg 2002). Furthermore, not every attempt to introduce a tradable permit approach has been successful. In air pollution control, attempts to establish transferable permit approaches have failed in Poland (Zylicz 1999) and Germany (Scharer 1999). The initial attempts to introduce a $SO_2$ trading system also failed in the United Kingdom (Sorrell 1999), although subsequent attempts to establish a $CO_2$ program in that country have succeeded.

On the other hand, it does appear that the introduction of new, tradable permit programs becomes easier with familiarity. In the United States following the very successful lead phase-out program, new supporters appeared and made it possible to secure passage of the Sulfur Allowance Program. The introduction of the various flexibility mechanisms (that involve different forms of emissions trading) into the Kyoto Protocol was facilitated by the successful experience with the U.S. Sulfur Allowance Program, among others. And the recent introduction of tradable permit systems in several European countries and the European Union was precipitated by the opportunities provided by the Kyoto Protocol.

It also seems clear that, to date at least, using a free distribution approach to the initial allocation has been a necessary ingredient in building the political support necessary to implement the approach (Raymond 2003). Existing emitters frequently have the power to block implementation, while potential future emitters do not. This has made it politically expedient to allocate a substantial part of the economic rent that is associated with these permits to existing users as the price of securing their support.

While this strategy reduces the adjustment costs (the cost of moving from one form of control to another) to existing emitters, it generally raises them for new users. In the typical U.S. case, new emitters must purchase all permits while existing emitters get allocations free-of-charge.[8]

Giving permits away is not without cost since it also tends to lower the efficiency of implementing emissions trading (Goulder et al. 1999). Auctioning the permits, instead of allocating them free-of-charge, allows the revenue to be used for reducing other distortionary taxes, producing a "double dividend."

One tendency that arises in some new applications of this concept is placing severe restrictions on its operation as a way to quell administrative fears about undesirable, unforeseen outcomes. The early Emissions Trading Program in the United States in the 1970s and early 1980s had this characteristic. For example, every created emissions reduction credit had to be certified, and all trades between noncontiguous sources had to be accompanied by air quality monitoring. In addition, banking was disallowed initially in the lead phase-out program. Although with increased familiarity (and comfort) the initial restrictions tend to disappear over time, extra constraints can diminish severely the early accomplishments of a program.

## *Environmental Effects*

One common belief about tradable permit programs is that their environmental effects are determined purely by the imposition of the emissions limit, an act that is considered to lie outside the system. Hence, it is believed, the main purpose of the system is to reduce costs, not to protect the air.

In retrospect, it is now clear that is an oversimplification for several reasons. First, whether it is politically possible to set an aggregate limit at all may be a function of the policy intended to achieve it. Second, both the magnitude of that limit and its evolution over time may be related to the policy instrument choice. Third, the choice of policy regime may affect the level of monitoring and enforcement and effects on compliance can affect outcomes. Fourth, the policy may trigger environmental effects that are not covered by the limit.

## The Stringency of the Limit

In general, the evidence seems to suggest that by lowering compliance costs, emissions trading facilitates the setting of more stringent caps. The lower costs offered by trading have been used in initial negotiations to secure more stringent pollution control targets (e.g., acid rain, lead phase-out, and RECLAIM programs) or earlier deadlines (e.g., lead-phase out program). The air quality effects from more stringent limits were reinforced by the use of adjusted offset ratios for trades in nonattainment areas. (Offset ratios were required to be greater than 1, implying a portion of each permit acquisition would be retired and result in improved air quality.) In addition, environmental groups have been allowed to purchase and retire allowances (e.g., acid rain program). Retired allowances represent pollution that is authorized but not emitted.

## Meeting and Enforcing the Limit

Regardless of how well any tradable permit system is designed, noncompliance can prevent the attainment of its economic, social, and environmental objectives. Although it is true that any management regime faces monitoring and enforcement issues, emissions trading raises some special issues. One of the most desirable aspects of emissions trading for emitters, their ability to increase profits, is a double-edged sword because it also can increase incentives for noncompliance when enforcement is lax or incomplete. In that case, higher profitability could promote illegal activity. Insufficient monitoring and enforcement also could result in the failure to keep a tradable permit system within its environmental limit.

In theory, the flexibility offered by tradable permit programs makes it easier to reach the limit, suggesting the possibility that the limit may be met more often under a tradable permit system than under the system that preceded it. That has been the case with fisheries (Tietenberg 2002), with some exceptions, but the evidence is less clear-cut for emissions trading.

The one case that obviously follows theoretical expectations is the Sulfur Allowance Program. For that program, compliance rates have been virtually 100%, due mostly to the requirement imposed at the time the program was implemented for continuous emissions monitoring (CEM). CEM makes noncompliance very easy to detect. Coupled with severe penalties for noncompliance, this approach clearly has led to improved compliance.

One increasingly important aspect of transferable permit systems in general, and emissions trading in particular, involves their ability to raise revenue for both enforcement and administration. In many permit systems, enforcement costs are routinely financed from the reduced costs or enhanced profitability promoted by the tradable permit system. Interestingly, this is rarely the case for air pollution, though it is now relatively common in fisheries (e.g., New Zealand).

The notion that emitters should finance at least some of the monitoring and enforcement costs is beginning to affect air pollution. In addition to the sulfur allowance system's industry-financed CEM system mentioned above, in the Danish greenhouse gas permit system (Penderson 2003), which does not rely on continuous emission monitoring, the electricity producers pay an administration fee of 0.079 DKK per ton of $CO_2$ allowance to the authorities. This covers the administrative costs associated with verification of $CO_2$ emissions, the control and distribution of allowances, operating the registry, monitoring of trading, and the development of the emissions trading scheme, among other costs.

## Effects on Air Quality

Many emissions trading programs are targeted at emissions reductions, not merely emissions stabilization. The U.S. programs to phase-out lead (Nussbaum 1992) and to reduce ozone-depleting gases (Hahn and McGartland 1989) have eliminated, not merely reduced, targeted pollutants. And in the case of lead, the program apparently was instrumental in achieving that reduction much more quickly than otherwise would have been possible (Nussbaum 1992).

Ex post data reveal that emissions have fallen (in some cases dramatically) following the introduction of emissions trading programs:

- In the Sulfur Allowance Program, sulfur emissions dropped 32.5% by 2003.[9]
- According to the annual RECLAIM audit, by 2002, $NO_x$ emissions had dropped by 56.8% and $SO_x$ emissions by 39.5%.[10]
- During the first year of the $NO_x$ budget program, ozone season (May through September) $NO_x$ emissions from power plants and other large combustion sources were reduced by more than 30% from 2002 levels. These emissions reductions occurred despite an increase in heat input (a measure of power generation) at affected sources.
- In the Chicago volatile organics program, participating sources reduced volatile organic chemical (VOC) emissions by 38% compared to their allotted emissions in 2000, the first year of the program, and by 47% in 2001. Allotted emissions generally represent a 12% reduction from sources' baseline emissions of VOCs emitted in the mid-1990s.[11]
- In the Santiago program, aggregate emissions fell from 7,442.5 kg/day in 1993 to 1,636.6 kg/day in 1999 (Montero et al. 2002).

While the case for substantial emissions reductions after the institution of emissions trading programs is compelling, the case for emissions trading as being fully responsible for those reductions is not. In the case of the Sulfur

Allowance Program, counterfactual baselines make it clear that the reductions directly attributable to the program were substantial (Ellerman et al. 2000), but making that distinction seems to be the exception rather than the rule.

Unfortunately, little evidence is available on the size of the reductions that might have been achieved from command-and-control, making it difficult to isolate the unique contributions of emissions trading. In both the Chicago volatile organics program and the Santiago program, where this question has been investigated, it seems clear that emissions trading was not only not fully responsible, but it may have been a relatively minor player.

In Chicago, for example, as of 2004 traditional regulation apparently has been much more important than the emissions trading program in motivating the large reductions in volatile organic matter (Kosobud et al. 2004), and in Santiago, the new availability of imported natural gas from Argentina seems largely responsible (Montero et al. 2002).

On the other hand, cap-and-trade versions of emissions trading have been successful, in most cases for the first time, in imposing a cap on aggregate emissions from certain sectors. Caps represent a fundamental change from the traditional approach of limiting only emissions from each source; without caps, as the number of sources grows, emissions grow.

The air quality impact of a cap can be even more profound when it is placed on a previously unregulated pollutant. For example, analysis of future climate change outcomes points out that the existence of a cap on greenhouse gases can have a profound effect on aggregate emissions levels over time (Working Group III 2001).

## Examining the Leakage Problem

Even if all emissions trading reductions proved to be greater than otherwise would have been obtained, one further possibility would be that the emissions were relocated, rather than reduced. Leakage occurs when pressure on the regulated resource is diverted to an unregulated or lesser regulated resource. Polluters moving their polluting factories to countries with lower environmental standards would be one example of leakage.

According to the "pollution havens" hypothesis, polluters affected by stricter environmental regulations in one country could be expected to move their dirtiest production facilities to countries with less stringent environmental regulations (presumed to be lower income countries) or face a loss of market share triggered by higher costs. In addition, consumers in the country with the strict regulation have an incentive to prefer the cheaper goods produced in the pollution havens, leading to a "race to the bottom" in terms of environmental quality.

Pollution levels can change in the pollution havens for three different reasons: (1) the composition effect; (2) the technique effect; or (3) the scale effect.

According to the composition effect, emissions change as the mix of dirty and clean industries changes; as the ratio of dirty to clean industries increases, emissions increase, even if total output remains the same. (Notice that this is the expected outcome from the pollution havens hypothesis.) The technique effect involves the ratio of emissions per unit output in each industry. Emissions could increase in pollution havens if each firm in the pollution haven became dirtier as a result of the openness of its host country to trade. And finally the scale effect looks at the role of output levels on emissions; even if the composition and technique effects were zero, emissions could increase in pollution havens simply because output levels increased.

What is the evidence on the empirical validity of the pollution havens hypothesis and its "race to the bottom" implication? Earlier surveys of the empirical work, such as Dean (1992), find absolutely no support for the effect of environmental regulation on either trade or capital flows. Jaffe and Stavins (1995) reach the same conclusion in their survey of the effect of environmental regulations on U.S. competitiveness. Several recent studies reviewed by Copeland and Taylor (2004), however, find that environmental regulation can influence trade flows and plant location, all other things being equal, though the effects still seem to be small.

Studies that attempt to isolate composition, technique, and scale effects generally find that the composition effect (the most important effect for confirming the pollution havens hypothesis) is small relative to scale effects. Furthermore, technique effects normally result in less, not more, pollution (Mani and Wheeler 1998). Though pollution can be increased through the scale effect, these findings are quite different from what would be expected from a race to the bottom.

Because the Kyoto Protocol is incomplete in its coverage, the possible movement of firms from covered nations to nations that have no obligations has played a significant role in the implementation discussions. The results from one ex ante study (Babiker 2005) suggest that significant relocation of energy-intensive industries away from the Organisation for Economic Co-operation and Development countries could occur, depending on the type of market structure, with leakage rates as high as 130%, in which case greenhouse gas control policies in the industrialized countries could actually lead to higher global emissions.

Some analysis is emerging of strategies for alleviating leakage effects. Mæstad (2001) examined the efficiency of several specific strategies that might be employed by a covered country to protect itself from an exodus of firms due to climate change policy. The main findings of this study were that if localization subsidies are available, the use of fossil fuels should be taxed at the Pigouvian rate while incentives to stay in the home country should be implemented, partly through trade provisions on final goods and partly through a localization subsidy to domestic firms. The localization subsidy could be implemented either through a pure money transfer or through the provision of free emissions permits.

## Programmatic Influence on Organizational Structure

Although hard evidence on the point is scarce, a substantial amount of anecdotal evidence is emerging about how emissions trading can change the way environmental risk is treated within polluting firms (Hartridge 2003; McLean 2003; U.S. EPA 2002). This evidence suggests that environmental management used to be relegated to the tail-end of the decisionmaking process. Historically, the environmental risk manager was not involved in the most fundamental decisions about product design, production processes, or the selection of inputs, but rather was confronted with the decisions already made and told to keep the firm out of trouble. This particular organizational assignment of responsibilities inhibits the exploitation of one potentially important avenue of risk reduction: pollution prevention.

Because tradable permits put both a cap and a price on environmental risks, they tend to get corporate financial people involved. Furthermore, as the costs of compliance rise in general, environmental costs become worthy of more general scrutiny. Reducing environmental risk can become an important component of the bottom line. Given its anecdotal nature, the evidence on the extent of organizational changes that might be initiated by tradable permits should be treated more as a hypothesis to be tested than a firm result, but its potential importance is large.

## Comparing Ex Post with Ex Ante Outcomes

If we are largely dependent on ex ante studies, how reliable are they? As part of their comparison of experiences in Europe and the United States, Harrington et al. (2004) were able to compare ex ante and ex post estimates for the specific cases they studied. They found a reasonable degree of accuracy in the ex ante estimates. Interestingly, the cases in which emissions reductions were greater than expected involved incentive instruments (including but not limited to emissions trading). The cases in which reductions fell short of expectations involved regulatory approaches. Their findings, consistent with other literature, suggest that regulators may be unduly pessimistic about the performance of incentive instruments or unduly optimistic about the performance of regulatory approaches or perhaps both.

## Economic Effects

Few comprehensive ex post evaluation studies of emissions trading programs of any kind have been conducted. The only program that has been subjected to close scrutiny is the Sulfur Allowance Program.

Two detailed, ex post evaluations of the $SO_2$ allowance trading program have been conducted (Carlson et al. 2000; Ellerman et al. 2000). While both find that the program had achieved results at a significantly lower cost than expected, their interpretation of this finding differs. While Ellerman et al. find substantial savings due to the Sulfur Allowance Program, Carlson et al. find that the lower costs were due mainly to factors that they saw as exogenous to the program, such as a decline in the price of low-sulfur coal and improvements in technology that lowered the cost of fuel switching.

Though no ex post cost-benefit analyses of specific programs have been conducted, the U.S. government did conduct a major ex post cost-benefit analysis of all U.S. air programs during the period 1970–1990, a period in which both the Emissions Trading Program and the Lead Phase-out Program were in operation (U.S. EPA 1997). The analysis concluded that the air programs during this period resulted in a present value of benefits of approximately $22 trillion, while the present value of costs was estimated to be in the neighborhood of $0.5 trillion. Clearly, the benefits from regulating air pollution during this period dramatically exceeded costs.

While it is not possible to draw conclusions about the efficiency of specific air pollution programs (including emissions trading) from this study of all air pollution programs, a close look at the numbers reveals that phasing lead out of gasoline produced especially large net benefits. Lead was so damaging (and the effects so long lasting) that reducing exposure was obviously a very desirable policy; that program, of course, did involve emissions trading.

The evidence from RECLAIM is anecdotal but consistent with the evidence from the other programs. In an evaluation of RECLAIM, an EPA report relying on interviews with key participants found:

> Industry stakeholders believe that compliance costs have been reduced as a result of the program. Facilities were able to minimize costs by controlling emissions using the least costly methods and by altering the timing of control installations. Facilities are able to optimize their timing by replacing equipment or installing pollution control devices when these activities fit into manufacturing and production schedules. (U.S. EPA 2002, 50)

While the case for substantial cost savings from the sulfur allowance, the lead phase out, and the RECLAIM programs seems compelling, the same cannot be said for the early U.S. Emissions Trading Program (Dudek and Palmisano 1988; Hahn and Hester 1989; Tietenberg 1990). In that program, the vast majority of offset and bubble transactions involved sources under common ownership, rather than among different firms as anticipated by the pre-implementation analyses. Apparently, many potential opportunities to lower costs were not realized.

Understanding that unexpected outcome requires a deeper knowledge of how that program fit with traditional command-and-control approaches. Several characteristics of the Emissions Trading Program deviated from a cost-effective design, but two seem particularly important: (1) the design discouraged emissions trades between non-proximate sources by requiring case-by-case regulatory approval for all such trades and imposing a large burden of proof on those seeking to trade; and (2) mandating a series of specific technology requirements that could not be met with emissions reduction credits. In the former case, transactions costs were so high as to preclude many cost-effective trades, and in the latter case, some of the flexibility offered by emissions trading was eliminated by mandating particular technological options. Cost-effective alternatives to those technologies could not be exploited.

## Technical Change

Since the results from theory, reviewed in the previous chapter, provide mixed signals about the effectiveness of emissions trading in promoting technological progress, what can ex post studies contribute?

Quantitatively speaking, not much, it turns out, due to the paucity of relevant studies, but some useful insights do emerge. The evidence from their comparison of policies in Europe and the United States leads Harrington et al. (2004) to find general, although not universal, support for the view that market-based instruments provide greater incentives than regulation for continuing innovation over time.

Kerr and Newell (2003) examine technical change during the lead phase-out program. Employing a unique panel dataset on petroleum refineries covering the full period of the phase-out, they find that the adoption of pentane-hexane isomerization technology—a substitute for lead as a source of octane—was one of the major responses to the increased severity of regulation.

Their results further suggest that the tradable permit system was influential in encouraging the adoption of the new technology. In particular, the adoption propensity of refineries with low (relative to those with high) compliance costs was significantly greater under the tradable permit regime than would have occurred under a non-tradable performance standard. Kerr and Newell (2003) also find that increased regulatory stringency (the amount of lead allowed in gasoline for that year) encouraged greater adoption of lead-reducing technology.

Notice that this study examines the adoption of this technology rather than the commercialization of a new technology. What about evidence on the effect of emissions trading on the development of new technologies?

What about the Sulfur Allowance Program? In a detailed analysis of the evolution of sulfur oxide control from utilities in the United States, Taylor et al. (2005) conclude that the weight of evidence regarding innovation in $SO_2$ control technology does not support the superiority of the Sulfur Allowance Program as an inducement for environmental technological innovation as compared with traditional environmental policy approaches. Specifically they find:

> Repeated demand-pull instruments, in the form of national performance-based standards, along with technology-push efforts, via public RD&D funding and support for technology transfer, had already clearly facilitated the rapid maturation of wet FGD system technology that diffused from no market to about 110 Gwe capacity in twenty-five years. In addition, traditional environmental policy instruments had supported innovation in alternative technologies, such as dry FGD and sorbent injection systems, which the 1990 CAA provided a disincentive for, as they were not as cost-effective in meeting its provisions as low sulfur coal use combined with limited wet FGD application. (Taylor et al. 2005, *370*)

The interesting insight here is that emissions trading will promote cheaper approaches. When non-technology approaches (such as fuel switching) turn out to be cheaper, it is they, and not the new technology, that will be chosen.

What about the development of truly new technologies? In a revealing study, Popp (2003) examines the impact of the Sulfur Allowance Program from a different vantage point by combining data on steam-electric power plants with data on patents granted in the United States. His paper shows that implementing the Sulfur Allowance Program in 1990 did not necessarily lead to more innovation, as measured by patent counts, but did lead to more environmentally friendly innovation, as measured by the effect of the new innovations on the removal efficiency of new scrubbers.

Popp points out that before the Clean Air Act of 1990, plants were required to use the "best available technology" for pollution control, usually a scrubber. While this mandate did create incentives for innovations that would lower the costs of installing and operating scrubbers, it did not create any incentives for improving air quality by exceeding the standards. As a result, under the traditional approach improving the pollutant-removing efficiency of the scrubbers' was a low priority. Furthermore, mandating scrubbers provided little incentive for research and development on other pollution control methods.

## Transactions Costs and Administrative Costs

Economic theory treats markets as if they emerge spontaneously and universally whenever unmet needs create profitable opportunities. In practice, the applications examined in this review point out that participants frequently require some experience with the program before they fully understand (and behave effectively in) the market for permits. In the literature, this is known as the "learning by doing" effect.

For example, in the RECLAIM program, the pre-implementation analysis assumed that large facilities would over-control their emissions and sell their excess permits, thereby providing an adequate supply. However, initially most facilities installed controls, made process modifications, bought permits, or reduced production to stay in compliance. They did not go above and beyond what was required for compliance and did not focus on generating excess permits for revenue (U.S. EPA 2002).

This initial reluctance to trade seems to change with familiarity. Over the years, RECLAIM facilities apparently have become more familiar with and more efficient in buying and selling credits. In addition, instead of just helping facilities to buy and sell credits, brokers now discuss control options and other market opportunities with participants. As the price of permits has increased, the market has become more efficient because companies have invested the time and effort to understand this market and to use it to minimize compliance costs (U.S. EPA 2002).

Kerr and Maré (1999) estimate econometrically the effect of transactions costs on the cost-effectiveness of the U.S. lead phase-out program. Although they find that refineries generally trade efficiently, they also find evidence of transactions costs interfering with trading. Specifically, their results support the theoretical expectation that those refineries facing the highest transactions costs were less likely to trade.

Focusing on "first-trade" transactions costs, which are defined as the cost of making one trade rather than not trading at all, they find a loss on the order of 10–20% of potential gains from trade. A loss of cost-effectiveness comes not only from the failure to execute profitable trades but also from the dilution of value due to the transactions cost associated with each trade.

They also find consistent patterns of failure to trade. Refineries that were part of small companies, smaller refineries, and refineries that did not have other refineries to trade with within their company more frequently chose not to trade. Newer participants lack familiarity with the market and smaller participants may lack the resources to invest the time necessary to identify and execute cost-effective trades.

Programmatic design features also may affect transactions costs:

- Credit-based programs, such as the early U.S. Emissions Trading Program, typically involve a considerable amount of regulatory oversight at each step of the process (for example, certification of credits and approving each trade). In contrast, cap-and-trade systems rarely require either of the steps, instead using a system that compares actual and authorized emissions at the end of the year.

- Price transparency (making prices public) can reduce the uncertainty associated with trading and facilitate negotiations about price and quantity. One example is provided by the Chicago Board of Trade public auctions held each spring for the Sulfur Allowance Program. More recently, that program's seven-year forward auction provided some early evidence about the effects of proposed reductions of the $SO_2$ cap on the pricing of future vintage allowances.

- Organized exchanges (where buyers and sellers can meet) and knowledge-able brokers can lower the search costs for those seeking trades.

- Some experience derived from the RECLAIM program demonstrates that mechanisms for sharing information on available technologies can reduce duplication of efforts for larger firms and provide a larger menu of control options for smaller firms (U.S. EPA 2002).

Since design features vary so much from program type to program type, it is difficult to generalize insights about how administrative costs vary across programs.[12] Nonetheless, two themes that emerge are that the administration of emissions trading systems eventually involves fewer administrative person-hours (McLean 2003) and that the bureaucratic functions performed are quite different than under command-and-control approaches (McLean 2003; Harrison 2004).

The finding that fewer administrative hours are involved is not universal, and it typically results after some "learning by doing" has occurred. The EPA (2002, 55) evaluation of the RECLAIM program, for example, finds that shifting from command-and-control to a trading-based compliance system initially required a significant shift in resources and increased attention to compliance. Whereas monitoring in the traditional regulatory system often could be satisfied by a simple inspection to be sure the mandated equipment was up and running smoothly, emissions trading requires checking actual emissions against author-ized emissions (including those acquired through trade).

Focusing specifically on emissions, rather than on the use of a particular piece of machinery, requires increases in both administrative resources (in the areas of compliance, inspections, and audits) and emitter resources over and above investments in abatement (planning a compliance strategy, implement-ing the appropriate combination of abatement and acquiring permits, monitoring emissions, and reporting compliance) (Harrington et al. 2004). It also, of course, should provide better control over emissions.

In addition to consequences for the amount of resources required, changing administrative functions also have implications for the nature of the skills required by administrators. Those who can monitor and enforce compliance replace engineers who seek to identify the correct control strategies for sources and to negotiate permit exemptions.[13]

Finally, one unexpected finding that has emerged from ex post evaluations of emissions trading systems is the degree to which the number of errors in pre-existing emissions registries are brought to light by the need to create accurate registries for emissions trading (Pendersen 2003; Montero 2002; and Hartridge 2003). Although inadequate inventories plague all quantity-based approaches, emissions trading seems particularly effective at bringing them to light and getting them corrected.

# Summary

• Two types of studies have been used to assess cost savings and air quality impacts: ex ante analyses, which depend on computer simulations, and ex post analyses, which examine the actual implementation experience.

• A substantial majority, though not all, of the large number of ex ante studies found the command-and-control outcome to be significantly more expensive than the least-cost outcome.

• Ex ante studies also have found that when the need for additional reduction is so severe that the control authority has no choice but to impose emissions standards that are close to the limit of technological feasibility, the immediate potential cost savings typically are very small, though over time those savings can increase as new technologies are introduced.

• Actual cost savings are likely to differ from the ex ante estimates of potential savings for three main reasons: (1) the cost of the least-cost allocation may be measured with some error; (2) the cost of the command-and-control allocation may be measured with some error; and (3) the implemented emissions trading program can differ considerably from the idealized programs modeled by these analyses.

• Until recently, it appeared that emissions trading was introduced only after more familiar systems had been tried and proved inadequate. It now appears that the introduction of new emissions trading programs has become easier as experience has been gained from implemented programs.

• To date, free distribution of permits (as opposed to auctioning them off) seems to be a key ingredient in the successful implementation of emissions trading programs.

• Whereas conventional wisdom holds that emissions trading affects costs, but not air quality, that seems at best an oversimplification. In retrospect, we now know that the feasibility, level, and enforcement of the emissions cap all can be affected by the introduction of emissions trading. In addition, emissions trading either may degrade or improve environmental quality by triggering effects that are not covered by the cap.

• In general, air quality has improved substantially under emissions trading, but with the exception of the Sulfur Allowance Program and the lead phase-out program (where the case seems clear), the degree to which credit for these reductions can be attributed to emissions trading, as opposed to exogenous factors or complementary policies, is limited.

• The air quality impacts under credit-type programs, such as the U.S. Emissions Trading Program, were less pronounced and achievements came more slowly than expected.

- Credit programs seem to be characterized by higher transactions costs and administrative costs than cap-and-trade programs.

- Program design features can lower both administrative costs and transactions costs. Transactions costs can be lowered by making permit transactions transparent, by the availability of exchanges and knowledgeable brokers, and by the sharing of information on the availability of cost-effective abatement technologies, while administrative costs can be lowered by continuous emissions monitoring and by software that streamlines monitoring and reporting.

- Both regulators and environmental managers of emissions sources have experienced considerable "learning by doing" effects, with the result that markets tend to operate much more smoothly after they have been in existence for a while.

- Although few detailed ex post studies have been accomplished, completed studies typically find that cost savings are considerable but less than would have been achieved if the final outcome were fully cost-effective.

- A cautionary note about interpreting these studies comes from the ex ante analyses that take into account the effects on other markets. When these effects are considered in the context of general equilibrium models, the results suggest not only that they can be quantitatively important but that they affect the conclusions about the desirability of emissions trading as a policy tool. In particular, they can in some cases cause emissions trading to be less efficient than command-and control. Existing ex post studies do not consider these general equilibrium effects.

- The literature contains some support for the finding that emissions trading encourages both emissions-reducing innovation and the adoption of new emissions-reducing technologies, but this is not a universal finding. In some circumstances, emissions trading may encourage the exploitation of low-cost strategies, such as fuel switching, that do not promote technical change in preference over the development of new technologies, thereby delaying their commercialization. The available evidence is too sparse to draw firm conclusions about which effect dominates.

## Notes

1. Specifically, the typical algorithm finds the source-specific emission reductions that solve equation sets (3)–(7) for uniformly mixed assimilative pollutants, equation sets

(15)–(19) for non-uniformly mixed assimilative pollutants, or equation sets (26)–(30) for uniformly mixed accumulative pollutants.

2. Russell (1981) has attempted to assess the importance of the facilitating role by simulating the effects on permit markets of regional economic growth, changing technology, and changing product mix. This study finds that for almost every decade and pollutant, a substantial number of emissions reduction credits would have been made available by plant closings, capacity contractions, product-mix changes, or the availability of new technologies. None of these opportunities for cost savings would have been picked up by a static model.

3. None of the studies from the recent OECD Workshop on Ex Post Evaluation (2004), for example, attempt this type of evaluation.

4. Interestingly, the ex post empirical studies have more to say about transactions costs than they do on market power. While many ex ante studies traditionally have focused on market power, the ex post studies cast more light on transactions costs. Our long history of modeling market power combined with the fact that suggestive data are available ex ante (i.e., number of players, market share, etc.) may bias the ex ante agenda toward the analysis of market power, while our theory about transactions cost is relatively less rich and the evidence of it usually emerges only once the market commences operation.

5. The recovery of monitoring and enforcement costs from users now has become standard practice in some other applications of the transferable permit system, such as fisheries (National Research Council Committee to Review Individual Fishing Quotas 1999). The Sulfur Allowance Program mandates continuous emissions monitoring financed by the emitting sources but does not tax sources to recover expenditures on other aspects of monitoring and enforcement activities.

6. See Spofford (1984, *30* and *51*).

7. Sulfates are not on the federal list of pollutants requiring state action. The decision to regulate sulfates and the choice of standards were the result of a unilateral California initiative.

8. As pointed out earlier, some new plants in the EU Emissions Trading Scheme receive gratis permits rather than having to buy them, so this point does not apply to that program.

9. http://www.epa.gov/airmarkets/cmprpt/arp03/so2_emissions_reductions.gif (accessed 2/11/05).

10. http://www.aqmd.gov/reclaim/index.htm (accessed 2/11/05).

11. http://www.epa.state.il.us/news-releases/2003/2003-048-erms.html.

12. Refusing to draw general conclusions, Harrington et al. (2004) suggest that whether administrative costs are higher or lower for emissions trading depends on the context and nature of the problem being addressed.

13. Recognizing this shift might prove to be a barrier to implementation if existing regulatory staffs worry about losing their jobs to those with skills more appropriate to managing emissions trading.

# 4

# The Spatial Dimension

Sulfur dioxide, particulates and, to some extent, nitrogen dioxide are appropriately classified as non-uniformly mixed assimilative pollutants.[1] If the resulting allocation of control responsibility for these pollutants is to be cost-effective, the theory reviewed in Chapter 2 is convincing on the need for control authorities to consider where pollutants are injected into the air as well as how much. Unfortunately, introducing source location into the policy design complicates matters; it is easier said than done. It is not surprising, therefore, that not a single operating air pollution emissions trading program follows the ambient permit model.

This chapter explores not only the difficulties of implementing a theoretically optimal policy but also the practical alternatives. Just how "bad" are policies that ignore spatial considerations when they clearly matter? How can modified emissions trading designs begin to forge a reasonable, if suboptimal, compromise between administrative complexity and reliable control?

## Difficulties in Implementing an Ambient Permit System

As noted in Chapter 2, the challenge posed by non-uniform mixing is easily manageable from a purely theoretical point of view. All the control authority has to do is to implement the ambient permit system described in that chapter. Unfortunately, the implementation of this system would not be a trivial matter. Both state control authorities and sources would have to overcome some rather formidable administrative and legislative barriers if an ambient permit system is to work smoothly in practice.

## Transaction Complexity

The first such barrier is the inherent complexity of an ambient permit system. Because typical laws mandate that the ambient standards be met everywhere, complete assurance that violations would not occur requires a very large number of receptor locations. Fortunately, complete assurance is not required.

Reasonable assurance can be gained with relatively few receptor locations. Because any particular flow of pollutants will affect a number of sites, the readings at contiguous monitoring sites are highly correlated. Due to this interdependence, a small number of carefully placed monitors can give an adequate picture of pollutant concentrations over a fairly large geographic area. The studies investigating this question typically find that nine or ten selected monitoring sites are adequate to cover a typical urban airshed.[2]

Though designing a transferable permit system to produce the desired concentrations at nine or ten sites is certainly more manageable than designing one for a much larger number of sites, it is a far from trivial exercise. To ensure that trades do not jeopardize attainment at any of these monitoring locations, a separate market is required for each monitored location.

The traded permits would have to be defined in terms of the reduction in concentration achieved at each of the nine or ten monitor locations. Each of these monitor-specific concentration permits could be traded independently of the others. Fewer markets would leave some monitors unprotected, raising the possibility that trades would trigger violations at one or more of them.

The complexity of this system is illustrated by a description of what any particular source would have to go through to negotiate a trade. Suppose a source wished to expand its production facility in a nonattainment area. It could legitimize the resulting increase in emissions by purchasing sufficient concentration permits from each affected monitor market. Because this proposed increase in emissions could be expected to affect concentrations at most, if not all, monitor locations, the source would be required to purchase a different number of offsetting permits in each of the markets. Each set of concentration permits would command a different price, reflecting the difficulty of meeting the ambient standard at the associated monitor.

Since the emission increase is not legitimized until all required offsetting credits are obtained, the expansion could be jeopardized by problems in any one of these markets. Problems could arise, for example, when few sellers exist in one or more of the markets. Markets with few sellers provide less assurance that competitive prices will prevail. When permit prices are not competitive, the transactions generally will not lead to a cost-effective allocation.[3]

Furthermore, prices may be more uncertain in markets with few participants; prices for ambient permits may be more commonly negotiated bilaterally on a case-by-case basis rather than determined by a market involving a large number of buyers and sellers. When a source is required to negotiate in more than one market (as it normally would be in an ambient permit approach), its problem becomes acute. The demand for credits in any one of the markets would depend not only on permit prices in that market but on the prices in all other markets as well. The source could not definitively negotiate in market A until it knew the price in market B and vice versa. The interdependency among these markets creates an indeterminacy that only can be resolved in general by negotiating simultaneously in two or more markets at once. Though not impossible, this is a difficult burden for the source to bear.

This problem is exacerbated when control technologies are capable of controlling more than one pollutant.[4] In this circumstance, the desired number of permits for one type of pollutant will depend on the number of credits obtained for the other pollutant and vice versa. Not only would the source be required to conduct simultaneous negotiations in different monitor markets of the same pollutant, it also would need to conduct simultaneous negotiations among the various markets associated with the different, but related, pollutants. The ability of sources to deal effectively with these interdependencies is questionable.

## Control Over Emissions

At several points in the Clean Air Act, as well as in various legal cases interpreting it, a reduction in emissions is explicitly stipulated as the main means of achieving the ambient standards. By creating nonattainment areas, Congress recognized and dealt with the obvious fact that these areas could not achieve compliance with the ambient standards for one or more pollutants by the statutory deadline. In the interim, as one condition of extending these deadlines, Congress required reasonable further progress toward meeting the standards on an annual basis. For our purposes, the important characteristic of the requirement for reasonable further progress is that it was defined in terms of emissions, not air quality.[5] Strategies that would improve air quality at specific locations in nonattainment areas are prohibited by this provision if they simultaneously allow any increase in aggregate emissions. In nonattainment areas it is not permissible to simply rearrange the location of emissions within the area to meet the standard, even when the standards could be met more rapidly and more cheaply this way.[6]

The reasonable further progress provision constrains and can rule out the ambient permit approach to pollution control. Although ambient permits capitalize on source location and dispersion to improve air quality at a significantly lower cost than is possible with other approaches, some trades would allow emissions to increase. When sources having a large impact on non-complying receptors (those recording air quality levels that violate the standards) sell permits to sources having an impact on receptors with significantly better-than-required air quality, emissions could increase. Despite the fact that these trades might reduce the cost of complying with the ambient standards, achieve more rapid compliance by concentrating the reductions on those monitors experiencing the violations, and lower the damage caused by the pollution by spatially diffusing emissions, they would be ruled out by the current statutory requirement for emissions reduction.

The available empirical evidence suggests that this is not a trivial point. All studies that the author is aware of that have examined this question find that emissions levels associated with the least-cost allocation are larger than command-and-control emissions levels. The results of those studies are presented in Table 4-1. Though the range is rather wide, it is clear that a substantial portion of the potential cost savings that conceivably could be achieved by an ambient permit system is directly attributable to the smaller emissions reductions needed to meet the ambient standards when location and dispersion are taken into account. In the United States, this portion of the cost savings is ruled out by current legislation.

While the ambient permit system may be a perfect theoretical solution to the problem of incorporating source location, in practice it is difficult to implement. Although specific legal barriers could fall before a congressional modification of the Clean Air Act, the administrative complexity of the system is inherent.

## Possible Alternatives

The vulnerability of the ambient permit system to these practical problems creates the need to examine administratively and legally feasible alternatives that are environmentally equivalent. Although these typically may not sustain the least-cost allocation, they hold out some potential for being less costly than the traditional approach. Several such approaches are considered here: (1) emission permit systems; (2) various types of zonal permit systems; (3) single-market ambient permit systems; and (4) trading rules.

### Emission Permits

One way to deal with the spatial complexity of pollution control is to ignore it. While from Chapter 2 it is clear that an emissions permit system would not support a cost-effective allocation of the control responsibility for non-uniformly mixed assimilative pollutants (only an ambient permit system could do that), the theory provides no evidence on just how large the potential cost penalty would be. If it were small, emissions permit systems might become an attractive second-best approach. Because they are administratively simple and by design ensure direct control over emissions, they avoid the two previously discussed inherent weaknesses of the ambient permit system.

**TABLE 4-1.** A Comparison of Emissions Reduction for Command-and-Control and Least-Cost Approaches

| Study and year | Pollutants considered | Geographic region | Command-and-control benchmark | Least-cost emissions reduction as a percentage of command-and-control emission reduction[a] (percent) |
|---|---|---|---|---|
| Atkinson–Tietenberg (1982)[b] | Particulates | St. Louis | State implementation plan requirements | 50.0 |
| Seskin et al. (1983) | Nitrogen dioxide | Chicago | Proposed RACT standards | 14.3 |
| Krupnick (1983) | Nitrogen dioxide | Baltimore | Proposed RACT standards | 51.6 |
| McGartland (1984) | Particulates | Baltimore | State implementation plan requirements | 92.2 |
| Spofford (1984)[c] | Particulates | Lower Delaware Valley | Equal percentage reduction | 83.0 |
| | Sulfur dioxide | | Equal percentage reduction | 73.3 |

Sources: Scott E. Atkinson and T.H. Tietenberg, "The Empirical Properties of Two Classes of Designs for Transferable Discharge Permit Markets," Journal of Environmental Economics and Management 9, no. 2 (1982): 116 (figure 5); Eugene P. Seskin, Robert J. Anderson, Jr., and Robert O. Reid, "An Empirical Analysis of Economic Strategies for Controlling Air Pollution," Journal of Environmental Economics and Management 10, no. 2 (1983): 117 (table 1); Alan J. Krupnick, "Costs of Alternative Policies for the Control of NO₂ in the Baltimore Region," unpublished Resources for the Future working paper (1983), 22 (table 4); Albert Mark McGartland, "Marketable Permit Systems for Air Pollution Control: An Empirical Study," unpublished Ph.D. dissertation, University of Maryland (1984), 67a (table 4.2); Walter O. Spofford, "Efficiency Properties of Alternative Source Control Policies for Meeting Ambient Air Quality Standards: An Empirical Application to the Lower Delaware Valley," unpublished Resources for the Future discussion paper D-118 (1984), 101 (table 21).

a. Each pair of allocations yields pollution levels which meet the same ambient standards.

b. The 12 g/m³ standard is used for this calculation.

c. Compares the least-cost strategy with the single-zone uniform percentage reduction.

The evidence on the size of the potential cost penalty when emissions permit systems are used to control non-uniformly mixed assimilative pollutants is presented in Table 4-2. The potential abatement costs of an emissions permit system are compared with those of the command-and-control approach and the ambient permit market (least-cost) allocations. In each study, all three allocations of pollution control responsibility are defined such that they meet comparable ambient air quality standards.

In the fifth column of Table 4-2, the potential abatement cost of the emissions permit system is compared with that of the traditional command-and-control approach. Because neither the emissions permit allocation nor the command-and-control allocation are least-cost allocations for non-uniformly mixed assimilative pollutants, the cheaper allocation can be only identified empirically. While an ambient permit system necessarily would have lower abatement costs, an emissions permit system may not. A ratio of greater than 1 indicates that the emissions permit approach achieves the objective at lower cost, while a ratio of less than 1 indicates that the traditional regulatory approach is cheaper.

Perhaps the most obvious characteristic of these data is the variability of the results among various pollutants and regions. The cost-effectiveness of an emissions permit system in this context apparently is quite sensitive to local conditions.

The difference in the cost of control resulting from the use of these two different approaches can be decomposed into two components: (1) the equal-marginal-cost component; and (2) the degree-of-required-control component. The equal-marginal-cost component refers to the amount of the difference due to the equalization of marginal costs of control that would occur with an emissions permit system but not with the command-and-control approach. For any comparable degree of required reduction, the emissions permit system would achieve that reduction at a lower cost. This component unambiguously favors the emissions permit system.

The second component derives from the fact that the degree of required emissions reduction is not usually the same for the two systems. Because the location of the sources matters, the degree of required emissions reduction depends on the allocation of control responsibility among sources. Since the two systems result in different allocations of control responsibility among sources, the total amount of emissions reduction needed to meet the ambient standards would not necessarily be the same.

The sign of this component is ambiguous. It can favor either the command-and-control system or the emissions permit system, depending on which requires more control. How much control each requires can be determined only in a specific context.

Though his study of the Santiago emissions trading system does not present differences in emissions (and therefore is not in Table 4-1), O'Ryan (1996) does present an interesting experiment. In one set of simulations, he makes policies comparable in terms of concentrations achieved at each receptor, not merely those where air quality exceeds the ambient standard. This eliminates the degree-of-required-control component as a source of cost advantage for the ambient permit system and is thus a test of how important the equal-marginal-cost component is. His results suggest that a substantial part of the cost advantage obtained from using an ambient permit system is due to the degree-of-required-control component (i.e., the ability to allow increased total emissions by strategically rearranging emission location).

In summary, whether the emissions permit or the command-and-control allocation is cheaper depends on the sign and magnitude of the degree-of-required-control component. If the command-and-control allocation requires more control, the emissions permit system unambiguously results in lower control costs; both the equal-marginal-cost and degree-of-required-control components act in the same direction, reinforcing one another. Whenever the emissions permit system requires more control, then the two components are of opposite sign and tend to offset each other. If the amount of reduction required in the emissions permit system is sufficiently large, the degree-of-required-control component would dominate the equal-marginal-cost component, causing the cost of control to be higher with an emissions permit system.

Table 4-2 is of some help in identifying those local conditions that affect the interaction of these two components. One of those conditions is the stringency of the ambient standard relative to the level of uncontrolled emissions. Generally, the higher the proportion of regional emissions that need to be controlled, the more the abatement costs of the two approaches converge. The convergence of these costs as the air quality standard becomes more stringent is indicated clearly, for example, in the Atkinson and Lewis (1974) study.

A second clue is provided by the studies in which an emissions permit system actually is more expensive than the traditional approach, a condition found in five of the twelve cases represented. (Though the dominance of the degree-of-required-control component was a clear theoretical possibility, it is rather striking how prevalent the phenomenon is, at least in these studies.) These studies are quite helpful in discerning the determinants of the degree-of-required-control. The degree of emissions reduction required by a permit system is quite sensitive to the spatial configuration of sources. When a few large sources are clustered near the receptor requiring the largest improvement in air quality, they would have to be controlled to a very high degree. Their resulting high marginal costs of control would be mirrored by equivalently high marginal costs of control for distant sources, despite the fact that emissions from distant sources have very little impact on the monitors where the greatest air quality improvement is needed. This over-control of distant sources results in much more emissions reduction than necessary to meet the ambient standard under a command-and-control approach.

**TABLE 4-2.** Using Emissions Permit Systems to Control Non-Uniformly Mixed Assimilative Pollutants: The Potential Cost

| Study and year | Pollutants covered | Geographic area | CAC benchmark | Ratio of CAC to EPS abatement cost | Ratio of EPS to APS abatement cost[a] |
|---|---|---|---|---|---|
| Atkinson and Lewis (1974) | Particulates | St. Louis metro area | SIP regulations | 6.00 | 1.67[b] |
| | | | | 1.33 | 4.51[c] |
| Roach et al. (1981) | Sulfur dioxide | Four Corners in Utah, Colorado, Arizona, and New Mexico | SIP regulations | 1.70 | 2.50 |
| Hahn and Noll (1982) | Sulfates | Los Angeles | California regulations | 1.05 | 1.07 |
| Atkinson (1983) | Sulfur dioxide | Cuyahoga County, Ohio | SIP regulations | 0.78 | 1.91[d] |
| | | | | 0.91 | 1.40[e] |
| McGartland (1984) | Particulates | Baltimore | SIP regulations | 2.50 | 1.88 |
| Krupnick (1983) | Nitrogen dioxide | Baltimore | Proposed RACT regulations | 0.69 | 8.64[f] |
| Seskin, Anderson, and Reid (1983) | Nitrogen dioxide | Chicago | Proposed RACT regulations | 0.42 | 33.9 |
| Spofford (1984) | Sulfur dioxide | Lower Delaware Valley | Equal percentage reduction | 0.83 | 21.3[g] |
| | Particulates | | | 11.10 | 1.97[h] |
| O'Ryan (1996) | Particulates | Santiago, Chile | Equal percentage reduction | 1.25[i] | 8.00[j] |

*Definitions*: RACT = Reasonably available control techniques; EPS = Emission permits system; SIP = State implementation plan; APS = Ambient permit system; CAC = Command-and-control.

*Sources*: Scott E. Atkinson and Donald H. Lewis, "A Cost-Effectiveness Analysis of Alternative Air Quality Control Strategies," *Journal of Environmental Economics and Management* 1, no. 3 (1974): 247; Fred Roach, Charles Kolstad, Allen V. Kneese, Richard Tobin, and Michael Williams, "Alternative Air Quality Policy Options in the Four Corners Region," *Southwestern Review* 1, no. 2 (1981): 44–45 (table 3); Robert W. Hahn and Roger G. Noll, "Designing a Market for Tradeable Emissions Permits," in Wesley A. Magat, ed., *Reform of Environmental Regulation* (Cambridge, MA: Ballinger, 1982), 132–133 (tables 7-5 and 7-6); Scott E. Atkinson, "Marketable Pollution Systems and Acid Rain Externalities," *Canadian Journal of Economics* 16, no. 4 (1983): 716 (table 4); Albert Mark McGartland, "Marketable Permit Systems for Air Pollution Control: An Empirical Study," unpublished Ph.D. dissertation, University of Maryland (1984), 67a and 77a (tables 4.2 and 5.2a); Alan J. Krupnick, "Costs of Alternative Policies for the Control of NO$_2$ in the Baltimore Region," unpublished Resources for the Future working paper (1983), 22 (table 4); Eugene P. Seskin, Robert J. Anderson, Jr., and Robert O. Reid, "An Empirical Analysis of Economic Strategies for Controlling Air Pollution," *Journal of Environmental Economics and Management* 10, no. 2 (1983): 117 and 120 (tables 1 and 2); Walter O. Spofford, Jr., "Efficiency Properties of Alternative Source Control Policies for Meeting Ambient Air Quality Standards: An Empirical Application to the Lower Delaware Valley," unpublished Resources for the Future discussion paper D-118 (1984), 47 and 50 (tables 7 and 8); R. O'Ryan, "Cost-Effective Policies to Improve Urban Air Quality in Santiago, Chile," *Journal of Environmental Economics and Management* 31, no. 3 (1996): 302–313.

a. These columns assume emissions are reduced sufficiently by both policies to meet the ambient standards at all receptors. The ambient permit allocation is assumed to be identical to the least cost allocation.

b. Assumes air quality of 60 g/m$^3$ at worst receptor.

c. Assumes air quality of 40 g/m$^3$ at worst receptor.

d. Assumes emission reduction sufficient to meet local ambient standards.

e. Assumes emission reduction sufficient to meet local and long-range transport standards.

f. Uses 100 g/m$^3$ for EPS and 98 g/m$^3$ for APS.

g. Assumes air quality of 250 g/m$^3$ at worst receptor.

h. Assumes air quality of 80 g/m$^3$ at worst receptor and both point and area sources controlled.

i. Assumes a 60% reduction.

j. Assumes air quality of 75 g/m$^3$ at worst receptor.

Other spatial configurations of sources require less over-control of distant sources in an emissions permit system. When sources are more ubiquitous and no cluster dominates the most polluted receptor, a permit system would be able to achieve more balance between distant and proximate sources. In this circumstance, the air quality could be brought to the standard with both lower control costs and less total emissions reduction.

It is possible, at least crudely, to test the hypothesis that the relative cost advantage (or disadvantage) of the emissions permit system depends significantly on the amount of excess emissions reduction that is required. If this hypothesis has merit, we should expect to find a significant cost disadvantage for the emissions permit system in those studies where the amount of emissions reduction required by the command-and-control approach is less than that required by the emissions permit system.

Suppose that the various studies were ranked by the ratio of allowed permit emissions to allowed command-and-control emissions and these rankings were compared with the cost ratios. Table 4-3 presents the relevant ranks for all studies allowing the computation to be made.[7]

The first noticeable aspect of these data is that of the five studies where the emissions permit system abatement costs were higher than the command-and-control approach (ranks 1 through 5 in the last column), all require larger emissions reductions than the command-and-control allocation. (Larger emissions reductions are recorded as a number less than one in the first column.) In fact, larger emissions reductions for the permit system are not uncommon; they occurred in eight out of the ten air pollution studies listed. The degree-of-required-control component apparently commonly favors the command-and-control approach.

It is possible to measure the correlation between these rankings using the Spearman's rank correlation coefficient. If the hypothesis is valid, there should be a statistically significant positive correlation between this ranking and a ranking of the estimates based on the ratio of command-and-control to emissions permit abatement costs. If the hypothesis is not valid, the correlation should be either zero or negative. Since the Spearman rank coefficient is 0.87, the hypothesis that the true correlation is zero can be rejected with a 95% degree of confidence.

This result suggests that the amount of emissions reduction required is a significant factor in explaining the cost-effectiveness of the permit system. The fact that the correlation is not perfect, however, should not be overlooked, since this lack of perfection serves as a useful reminder that the test is a relatively crude one and that heterogeneity of control costs also is important.

Notice in Table 4-3 that two studies find that an emissions permit system requires less emissions reduction. As expected, they find that the control costs are lower for an emissions permit system since the two components reinforce one another.

**TABLE 4-3.** Correlation of Ranks Between Relative Level of Emissions Reduction and Control Cost: Command-and-Control and Emissions Permit Systems

| Study | Ratio of EPS emissions to CAC emissions | Rank | Ratio of CAC to EPS abatement cost [a] | Rank |
|---|---|---|---|---|
| Seskin, Anderson, and Reid (1983) | 0.21 | 1 | 0.42 | 1 |
| Atkinson (1983) | 0.32 | 2 | 0.78 | 3 |
| | 0.39 | 3 | 0.91 | 5 |
| Atkinson and Lewis (1974) | 0.50 | 4 | 1.33 | 6 |
| Krupnick (1983) | 0.71 | 5.5 | 0.69 | 2 |
| Roach et al. (1981) | 0.71 | 5.5 | 1.70 | 7 |
| Spofford (1984) | 0.74 | 7 | 0.83 | 4 |
| | 0.84 | 8 | 11.10 | 10 |
| McGartland (1984) | 1.09 | 9 | 2.50 | 8 |
| Atkinson and Lewis (1974) | 1.11 | 10 | 6.00 | 9 |

*Notes:* EPS = emission permits system; CAC = command-and-control.
*Source:* See Table 4.2.
a. This column is taken from Table 4.2.

Also of interest are the five studies (ranks 6 through 10 in column 5) that show lower control costs for the emissions permit system. For three of these, the emissions permit system was cheaper despite requiring larger reductions. The equal-marginal-cost and degree-of-required-control components for these three studies tended to offset each other, but the dominance of the equal-marginal-cost component caused the emissions permit system to produce a lower control cost.

Neither the command-and-control nor the emissions permit policy considers source location in assigning control responsibility. The magnitude of the cost penalty associated with ignoring source location in emissions reduction credit trades can be ascertained by comparing the emissions permit and ambient permit abatement costs, as is done in column 6 of Table 4-2. It is easy to see that for every one of these studies, except the one by Hahn and Noll (1982), location matters. The cost savings lost by ignoring source location are large, even for those pollutants and regions where the emissions permit system would be more expensive than the command-and-control allocation. By targeting the emissions reduction to those sources having the largest impact on the binding receptors, less reduction in total emissions is required.

The increase in allowable emissions that normally accompanies policies incorporating source location can be especially troublesome for pollutants that can be transported long distances. Ozone, sulfur oxides, and nitrogen oxides all fit in this category. For these pollutants, the computer-simulated cost savings may be misleading to the extent that they are based purely on the cost of meeting local receptors, allowing more emissions to be transported to other regions.

Atkinson (1983) investigated the significance of this potential bias by comparing the cost savings attributable to incorporating location when only local receptors were considered to that when the contribution of emissions to long-range transport also was considered.

His results suggest that the inclusion of long-range transport has two main effects: (1) it requires more total emissions reduction; and (2) it requires relatively more reduction from sources with tall stacks, since tall stacks enhance long-range transport. Atkinson's results indicate that although consideration of long-range transport tends to diminish the cost penalty associated with an emissions permit system (by requiring larger emissions reductions), it does not eliminate it. Even for long-range transport pollutants, the permit system still over-controls emissions; location still matters, though its influence is less significant than when only local receptors are considered.

Its normally large cost penalty is not the only strike against using an emissions permit system to control non-uniformly mixed assimilative pollutants. "Hot spots" are another. Hot spots arise when emissions are geographically concentrated, causing unacceptable ambient concentrations in that location.

By ignoring the location of the discharge point (the source of its simplicity), the emissions permit policy forces the control authority to relinquish control over concentrations. Though it can control the total level of emissions, unfortunately the correspondence between the total level of emissions and air quality measured at the monitors is not unique. Concentration levels are sensitive to the location as well as the amount of emissions. Emissions permits (as opposed to ambient permits) control only the latter.

The problem arises when the control authority attempts to issue the correct number of emissions permits to achieve an ambient standard. If it were perfectly omniscient, with full knowledge of the control costs of all emitters, defining the cap would be a simple matter. With this detailed control cost information, it would be possible to anticipate how the permits would be traded in a market even before the market was initiated. Combining its presumed knowledge of control costs with its presumed knowledge of all transfer coefficients, the authority could define a cap that would just meet the ambient standard at the worst receptor. The calculated cost penalties in Table 4-2 are based on just such an assumption about the behavior of this omniscient control authority.

But is that realistic? A truly omniscient control authority would not need permit markets to achieve cost-effectiveness. It could mandate cost-effective emissions standards for all sources directly without the bother of initiating permit markets. Indeed, it was the absence of this very information that triggered interest in permit markets in the first place.

What is likely to happen in practice? Because the control authority normally would not know which sources would end up trading the permits, it would in all probability build a safety margin into its calculation of the control baseline. By forcing more control than necessary to meet the ambient standards under conditions of perfect information, it could lower the likelihood of hot spots. Due to the need for this safety margin, the actual cost penalty associated with the emissions permit system would be larger under realistic assumptions about control authority behavior than modeled in the simulation studies.

Even in those cases where the cost penalty may not be larger than estimated, the risk of violating the ambient standards at one or more locations is increased by trading activity. Suppose, for example, that a particular allocation of pre-trade control responsibility is consistent with the ambient standards. For that moment, the control authority would have fulfilled its statutory obligations. However, as the number and composition of sources changed over time, permits would be traded and this assignment of control responsibility would be rearranged. Any rearrangement involving an increase in the number of permits held by those sources near binding receptors would jeopardize compliance. Nothing in the design of the emissions permit system prevents these concentrations from exceeding the ambient standard.

A third strike against an emissions permit approach stems from its inability to affect the location of new emissions sources. Since prices for emissions permits do not vary with location, the cost of pollution control for any potential emissions source controlled by that permit market would not depend on location either. Yet if that area is to meet the ambient standards over time, source location may be crucial. When a clustering of sources triggers nonattainment, for example, the program would have to be modified. In a nutshell, the emissions permit affords too little protection to the ambient standards over the long run by sending the wrong signals to potential polluters making location decisions.

Do emitters respond to these signals? Not much empirical work has been done on this, though one piece of research is relevant. Henderson (1996) finds that environmental regulations have caused polluting industries to spread out in the United States. Although environmental factors rank somewhat lower in location decisions than such traditional factors as access to markets and production, transport costs, and the characteristics of the site, for heavy emitters environmental control costs can be quite important. This evidence implies that the presence of location incentives in the ambient permit system could prove important in preventing clustering. Their absence (as in an emissions permit system) might allow clustering over time but at least it would not promote it.

Concern over hot spots has been important in the many emissions trading debates, including the Sulfur Allowance Program. In the pre-implementation discussions laying the groundwork for the Sulfur Allowance Program, the Northeast, which is widely thought to receive pollution emitted by power plants in the Midwest, was particularly concerned that trading might intensify the problem. Despite this concern, the Sulfur Allowance Program allows national trading on a one-for-one basis. Hence, it provides an interesting test case of whether trading results in hot spots.

What does the evidence reveal? Were these fears on target? Burtraw and Mansur (1999) examine how trading would affect geographic shifts in emissions. They find that trading would result in a sizable geographic shift in emissions, but the shifts would not adversely affect the Northeast. Allowing trading actually resulted in pollutant concentration decreases rather than increases in the East and Northeast. Deposition of sulfur in the eastern regions also decreased by a slight amount as a result of trading.

This work suggests that although emissions permit systems can in principle allow hot spots, that is certainly not automatic, and it may not even be the normal outcome. It depends on the location patterns of emitters and how those patterns evolve over time. With the Sulfur Allowance Program to date, emissions clustering has not been much of a problem, but it is impossible to know the extent to which that result would generalize to other pollutants and other geographic areas.

One final question that emerges for emissions permit systems used to meet ambient standards is the appropriate size of the trading area. Small areas can raise costs by ruling out cost-saving trades (those involving potential trading partners outside the trading area). Large areas, on the other hand, open the possibility of excessive control of distant sources—those that have little affect on the ambient air quality at the most affected receptors. The fact that costs rise when the trading area is either too small or too large raises the possibility of an optimal size for the trading area.

Atkinson and Morton (2004) investigate this question both theoretically and empirically. First they set out to derive a rule-of-thumb to guide decisions about expansions of trading areas. They show that it is possible to derive such a rule but, unfortunately, only for a highly stylized case with unrealistic assumptions.[8]

Reacting to these unrealistic assumptions, they investigate the question empirically in the context of a computer simulation of a program to meet nitrogen-loading standards for the Chesapeake Bay. By hypothetically increasing the geographic size of the emissions trading region, they are able to determine whether control costs increase or decrease for the emissions permit system given different levels of allowable loadings. They compare emissions permit control costs with those for an ambient permit system, whose total costs of control necessarily decline as the size of the trading region increases.

Their results indicate that the total cost difference depends dramatically on the required level of ambient improvement. For very stringent ambient standards, regional expansion of the size of the trading areas substantially reduced total control costs. As the standard became weaker, the cost reduction from expansion became smaller, until for much less-stringent ambient standards, regional expansion substantially increased the total costs of the emissions permit system.

One solution to the related problems of hot spots and over-control in an emissions permit system is to add some kind of constraint on the pure emissions permit system. In the United States, this problem has been attacked by "regulatory tiering," which implies applying more than one regulatory regime at a time.

In the Sulfur Allowance Program, sulfur emissions are controlled not only by the Sulfur Allowance Program but by the regulations designed to achieve local ambient air quality standards as well. All transactions have to satisfy both programs. Thus, trading is not restricted by spatial considerations (national trades are possible), but the use of acquired allowances is subject to local regulations protecting the ambient standards.

The second regulatory tier (the ambient standards) protects against hot spots (by disallowing any specific trades that would create them), while the first tier (the Sulfur Allowance Program) allows unrestricted trading of allowances. Because the reductions in sulfur are so large and most local ambient standards are not likely to be jeopardized by trades, this second tier is not expected to constrain very many, if any, trades. Yet its very existence offers sufficient assurance that local air quality will be protected to enable an emissions permit system to be implemented. This approach is analyzed in more detail in the "Trading Rules" section below.

## Zonal Permit Systems

Another possible approach considered from time to time is a zoned emissions permit system. In this approach, the control region is divided up into a specific number of zones, with each zone allocated a zonal cap. In pure zonal systems, permits can be traded within each zone on a one-for-one basis, but trading among zones is prohibited.

This system has a certain surface appeal because it appears to respond to some of the problems that plague both the emissions and ambient permit systems:

- Whereas the emissions permit system over-controls distant sources, the zoned permit system creates separate markets for distant and proximate sources.

- Whereas the emissions permit system is vulnerable to the creation of hot spots, the zoned permit system attempts to lower this vulnerability by reducing the number of trades between non-proximate sources.

- Whereas the ambient permit system requires each source to purchase permits from many markets, the zoned permit approach allows sources to operate in only one market.

In a crude way, the creation of zones takes location into account. The necessity for ambient modeling is eliminated by restricting trades to sources within the same proximate area. As long as all sources within each zone are closely clustered and stack heights are controlled (two very strong assumptions), all sources within each zone might be expected to have similar transfer coefficients. As long as the trading sources have similar transfer coefficients, emissions trades would not cause large changes in concentration at the relevant receptors.

Unfortunately, this historic rationale is flawed on a number of grounds, some of which can be identified even before we turn to the empirical evidence. The inability of sources to trade permits across zonal boundaries restricts trading opportunities and reduces the potential for cost savings. This inverse relationship between cost savings and zone size sets up an inherent conflict in determining the appropriate zone size. To provide maximum protection against hot spots, the zones should be relatively small. On the other hand, by restricting trading opportunities, small zones raise costs.

Williams (2003) models these tradeoffs explicitly and finds that the optimal number of zones will tend to increase with the degree to which the effects of pollution are localized. The more localized the effects of pollution, the greater the potential for hot spots, which will tend to favor a system with smaller trading zones. The slopes of the marginal benefit and marginal cost curves also matter, however, because they affect the cost savings from trading—steep marginal benefits favor smaller zones, while steep marginal costs favor larger zones.

He also notes that pollution sources generally should be placed in zones with other sources that are close substitutes for them (i.e., have similar transfer coefficients). Among other things, this important point requires consideration of the height of the discharge point in addition to spatial coordinates, since a source with a tall stack can have different effects than an otherwise similar source.

In practice, the implementation of a zonal permit system places a larger burden on the control authority than the implementation of an emissions permit system. With the zonal permit system, the control authority has to define the correct set of zonal caps, whereas an emissions permit system has to define only one. Because permits cannot flow across zonal boundaries, these administratively determined, initial caps set definite, permanent limits on the trading possibilities. Determining how much emissions reduction to assign to each zone is a crucial responsibility of the control authority in evaluating the cost-effectiveness of this approach.

In principle, at any point in time some specific allocation among zones minimizes the cost for a given configuration of zones.[9] However, to define that allocation the control authority would have to know the control cost functions of every source. Because such omniscience is an unrealistic expectation for any control authority, zonal allocations will in practice deviate from these full-information allocations.

Unfortunately, the more realistic, limited-information allocations would extract an additional cost penalty. Allocating too much control responsibility to one zone and too little to another would raise compliance costs above the least-cost solution even if the control authority were able to decide the correct total emissions reduction for the region as a whole.

Even if the control authority were able somehow to make the cost-minimizing assignment of caps among zones for a particular point in time, the normal evolution of the local economy and the composition of emitters would require changes in this assignment over time.

This discussion has suggested two sources of a cost penalty in the design of zonal permit systems: (1) the administrative allocation of caps to zones (coupled with the inability to correct this allocation by means of interzonal trades); and (2) the use of emissions (rather than concentration) reduction trades within zones. It also has suggested the need to assess the extent of the hot spot problem when the baseline control level is not sufficiently stringent to protect against violations of the ambient standards under all possible trades. Simulation models can add further clarification by calculating the magnitudes of the cost penalties and determining the seriousness of the hot spot problem.

## Full-Information Simulations

Full-information simulations presume omniscient control authorities. Both the least-cost total emissions reduction and the least-cost assignment of this reduction among zones, given the particular zonal configuration in that simulation, are assumed. Though unrealistic in their treatment of control authority behavior, these studies do tend to show the potential for zonal systems under the most congenial circumstances. These serve as a benchmark for our subsequent discussions of limited-information zonal permit systems.

In this full-information approach, as the number of zones is increased (by reducing the size of each zone), the cost-effectiveness of the policy must increase. Smaller zones not only mean less within-zone cost penalty, but the between-zone cost penalty is eliminated by the full-information assumption as well. Because the correct zonal caps are assumed, the need to trade credits across zone boundaries to reduce costs is eliminated.

Because each zone in this system represents an independent, functioning market, the interesting empirical question is how sensitive the remaining cost penalty is to the size of the market. The first study (Roach et al. 1981) to attack this question examined the effects of applying an emissions permit system on a regional, state, or airshed level, while another study by McGartland (1984) examined the effects of creating multiple zones within an airshed. Together, these studies encompass a wide range of market sizes.

For large regions, the Roach et al. (1981, 44) study finds that large reductions in the cost penalty could be achieved by reducing the size of the zones. If a single emissions trading system were used for the entire multi-state Four Corners region, for example, the control cost was estimated to be three to four times higher than if separate markets were created for each of the region's airsheds.

TABLE 4-4. The Effect of the Number of Zones on the Potential Cost-Effectiveness of a
Full-Information Zonal Permit Policy: Particulate Control in Baltimore, Maryland

| Number of zones | Annual potential control cost (millions of 1980 dollars) | Percent of cost penalty remaining |
|---|---|---|
| One | 66.82 | 100.0 |
| Three | 59.94 | 74.9 |
| Six | 48.42 | 32.7 |
| Nine | 42.97 | 12.7 |
| Fifteen | 42.03 | 9.3 |
| Least Cost | 39.49 | 0.0 |

*Source:* Albert Mark McGartland, "Marketable Permit Systems for Air Pollution Control: An
Empirical Study," unpublished Ph.D. dissertation, University of Maryland (1984), 77a (table 5.2a).

The higher cost of the single, large market is caused by the over-control of distant sources. To ensure compliance with the ambient standards in all locations,
larger regional emissions reductions are required. With multiple zones, the
reductions can be selectively concentrated on those zones where they are most
needed; targeting the reductions reduces the costs.

As shown in Table 4-4, the McGartland study finds further gains from multiple zones even within a single airshed. According to this study, it takes at least
three, and possibly as many as six, zones to cut the cost penalty in half. This is
a discouraging finding because the larger the number of zones, the more
restricted the trading opportunities among sources.

Though this is not an important restriction when the administrative allocations of caps among zones are optimal, it is crucial in the more realistic setting
of limited information. This point comes through clearly from studies that
have considered more realistic, limited-information zonal systems.[10] Rather
than use optimal conditions from the model to allocate zonal caps, these studies base these determinations on rules of thumb frequently used by control
authorities.

## *Limited Information Simulations*

Contrary to the expectation that small zone sizes would afford better control
over concentrations, these authors find that the hot spot problem could be
severe even with very small zones.[11] Close inspection of the results indicates that
different stack heights among sources within the same zone are a major reason
for this discrepancy. When within-zone stack heights vary considerably, even
contiguous sources may have very different transfer coefficients. Within-zone
emissions trades among sources with different transfer coefficients could produce drastic changes in locally measured concentrations. To avoid hot spots,

control authorities would have to increase the amount of required emissions reduction within each vulnerable zone to allow a margin of safety. This defeats one of the central purposes of a zonal permit system—the prevention of over-control.

These studies also discovered that the total cost of a limited information system would be sensitive to the initial zonal allocation of caps. Several rules of thumb that might be used by a control authority to make these zonal allocations were examined. They included: (1) equal percentage reductions based upon uncontrolled emissions; (2) equal percentage reductions in currently allowed emissions; and (3) reductions based on the need to improve air quality at the nearest within-zone receptor. All these had large cost penalties associated with them; none emerged as particularly superior or desirable. All conventional administrative approaches to allocating caps among zones seem to undermine the usefulness of the zonal permit approach.

Unfortunately, while these studies find that smaller zone sizes did not alleviate the hot spot problem, they also find that they significantly increased abatement costs.[12] In limited-information permit systems, the conventional initial allocations produce high cost penalties that can be reduced only by trading. Smaller zones restrict trading opportunities substantially (since no trades are permitted across zone borders) and the costs rise accordingly.

In realistic circumstances, zonal permit systems with no trading between zones do not appear promising. Because region-wide trades are an important source of cost reduction, the best systems must allow region-wide trades while not allowing hot spots to arise. The message from the simulation studies is that zonal permit systems do not in general provide much of an opportunity either to reduce costs or to control the hot spot problem.

## Zonal Systems with Directional Trading

One practice, used by the RECLAIM program, is a system of directional trading. In directional trading, limited zonal trading is allowed. Specifically, one zone is allowed to acquire permits from another zone but not vice versa.

The RECLAIM rules divide the trading area into two distinct zones: a coastal zone and an inland zone.[13] Since the inland zone is downwind of the coastal zone, emissions from the coastal zone can affect concentrations in the inland zone, but not vice versa. To prevent any trading-induced shifting of emissions from the inland zone to the coastal zone, sources on the coast are prohibited from acquiring emissions permits from inland sources, but inland sources can acquire permits from both coastal and inland sources.

The absence of full ex post evaluations for the RECLAIM system prevents us from knowing how much of the cost savings were foregone by this geographic restriction on trading. However, analysis of the geographical distribution of emissions during the first nine years of the program on a quarterly basis does not show any distinct shift in the geographical distribution of emissions (SCAQMD 2004). And the system could be expected to represent an improvement over a zonal system that allowed no trades between zones. Rather than eliminate all trades between zones, this system only eliminates the subset of trades that would be most likely to create hot spots.

## Zonal Permit Systems with Exchange Rates

Another possibility for improving the performance of zonal systems is developing a procedure for allowing trades between zones based upon an exchange rate. By exchange rate, we mean the ratio of the amount of emissions allowed in the buying zone for each unit of emissions reduced in the sellers zone. In an emissions permit system, of course, the exchange rate is 1 and in a pure zonal system with no interzonal trading it is 0, but allowing other ratios potentially could provide policymakers with an additional degree of freedom. How useful is the flexibility?

This question became an important one as Europe was seeking to implement the 1994 Second Sulfur Protocol. Researchers were investigating various emissions trading designs that might facilitate a cost-effective allocation of control responsibility.

In this setting, the hot spot problem was defined in a very specific way. Several areas were identified as especially sensitive to sulfur deposition, and "critical loads" were identified for those areas. The critical loads limited the amount of sulfur deposition in a specific area. The difficulty, of course, was that even if an initial allocation of control responsibility among countries respected those constraints, unconstrained trading would not.

The protocol assigned control responsibilities to countries, so the analysis took the form of asking whether, given that assignment, costs could be lowered by trading that respected the critical load constraints. Førsund and Naevdal (1998) examined the "existence" question on a theoretical level. Does there exist a predefined set of exchange rates such that the cost-effective solution would be reached by trading? In general, the answer is that no such set of exchange rates exists. Thus, any set of exchange rates must represent a second-best solution. Some exchange rates, however, are presumably better than others.

Klaassen and Førsund (1994) suggest that one promising candidate is setting the exchange rate equal to the ratio of weighted sums of transfer coefficients ($a_{ki}$) for all M receptors affected by each country. The weights would be the shadow prices ($\mu_k$) associated with each receptor.

$$\sum_{k=1}^{M} \mu k a_{ki}$$

The intuitive appeal of this exchange rate is that it takes into account not only the impact of the trade on each receptor (via the transfer coefficients) but also how difficult it is to meet the air quality constraint at each receptor (via the shadow prices). Unfortunately, as shown by Forsund and Naevdal (1998), in principle it is possible for exchange rate trading to lead to even higher costs than one-to-one trading.

How do exchange rates fare empirically? Forsund and Naevdal (1998) examine this question in the context of sulfur control in Europe. In general, they find that adding exchange rate trading to the initial allocation of responsibility would be capable of reducing abatement costs on the order of 19%.

That may be an overstatement. As shown by Hahn (1986) and Atkinson and Tietenberg (1991), if the trading is bilateral and sequential (as opposed to global and simultaneous), the outcome of a trading system is path dependent. Path dependence in this context means that the final equilibrium of the trading will depend on the trading sequence. Since the outcome will not be unique, it will be difficult to say just how cost-effective exchange rate trading will be without knowing the actual sequence of trades. And the actual sequence will not be known until those trades have been completed.

By empirically modeling various trading sequences, Atkinson and Tietenberg (1991) note that actual cost savings may fall short of cost savings estimated from models that assume global and simultaneous trading. Fortunately, as Burtraw et al. (1998) point out, the magnitude of the loss of cost savings due to bilateral trading is reduced if trades are allowed among individual sources rather than limited to countries, a point of particular relevance to the Kyoto Protocol.

As noted above, in principle it is possible for exchange rate trading to lead to even higher costs than one-to-one trading. Montero (2001) has investigated the conditions under which it is better to disallow interzonal trading than to use exchange rates. This study shows that the regulator should have pollution markets integrated through optimal exchange rates when the marginal-abatement cost curves in the different markets are steeper than the marginal-benefit curves; otherwise the markets should be kept separated.

The intuition behind this result is straightforward. Interzonal trading provides more flexibility to firms in case costs are higher than expected, but at the same time, it makes the amount of control in each market more uncertain. If the marginal-cost curves are steeper than the marginal-benefit curves, the regulator should pay more attention to the cost of control than to the amount of control and, therefore, have markets integrated. On the other hand, if the marginal-benefit curves are steeper than the marginal-cost curves, the regulator should pay more attention to the amount of control in each market and, therefore, have markets separated.[14]

## Single-Market Ambient Permit Systems

One possible means of allowing region-wide trades while providing some protection against hot spots involves using a bare-bones version of the ambient permit system. In this version, all trades are consummated on the basis of their effect on a single, worst-case receptor. Other receptors are presumed to benefit indirectly from the emissions reduction needed to meet the ambient standard at this receptor. The problem of transaction complexity is avoided in this version since there is only one trading market for concentration permits. Because region-wide trades are permitted, the cost penalty associated with restricted trading opportunities is avoided.

The degree of protection against hot spots afforded by this approach depends on local circumstances. Because the concentration permits are defined in terms of a single receptor, the associated receptor is adequately protected, but the others are protected only indirectly. As long as the air quality in a particular region is dominated by a single receptor, this approach typically would impose only a small associated cost penalty in the short run, while offering a substantial reduction in the complexity of compliance. The indirect protection would be adequate.

Adequate protection from hot spots, however, completely hinges on the dominance of the single receptor. Single-receptor dominance can be tested using the simulation models by examining the number of binding air quality constraints in the ambient permit solution. If single-receptor dominance exists, the air quality constraint will be binding at only one receptor. The air quality at the other receptors would be better than required. The price for the concentration reduction credit associated with the binding receptor would be positive, while the prices of the credits associated with the other receptors would be zero, reflecting their excess supply.

In general, the air simulation models find from one to three binding receptors. In the short run, the single-permit market would be fully cost effective in those cases with only one binding receptor; it would not be in the others. Therefore, more total emissions reduction would be required to guarantee compliance in air sheds with multiple binding receptors. This additional control adds to the cost penalty. The only study of how large the cost penalty would be in cases with multiple binding receptors (Atkinson and Tietenberg 1982, *114*) concludes that it remains very small. In this study, at least one receptor could serve as a useful proxy for the others, given the current configuration of sources.

This optimistic finding that a single market for concentration reduction credits would suffice is tempered to a considerable degree by two considerations. First, a single study is hardly conclusive. In the case of multiple binding receptors, the size of the amount of excess emissions reduction needed to ensure compliance would depend on the proximity of the binding receptors to each other, which would vary from airshed to airshed. Contiguous binding receptors would have lower cost penalties because one receptor could act as a reasonable proxy for the others. When the binding receptors are far apart, however, further reductions in concentrations at one will have only a minimal impact on the others. Therefore, a policy geared toward one receptor could not be expected to deal effectively with the concentration levels at more distant receptors.

The second consideration is even more troublesome because it is an inherent flaw and not one that depends simply on local conditions such as the number of binding receptors. A single market for concentration permits would establish locational incentives that eventually would undermine any claim to reasonableness, even in those areas initially having only one binding receptor.

With a single market for concentration permits, permits purchased for use at some distance from the receptor cost less per unit of emissions than credits for use near the receptor. For new sources, this creates an incentive to locate operations away from the initially binding receptor because credits needed to legitimize emissions would be cheaper there. When sufficient new sources had located near some new, previously remote receptor, that receptor would become binding as well, but the existing single-permit system would no longer be sufficient. Because this trading system is based on preventing violations only at the initially binding receptor, it creates not only the opportunity but the incentive for other receptors to be violated as well.

Should a second receptor become binding, it would, of course, be possible to initiate another market. Though two markets are still preferable to the nine or ten markets (one for each receptor) required by the full ambient permit system, the apparent simplicity advantage diminishes over time. As more and more receptors become binding, more separate markets would have to be initiated. As time passes, the complexity of this version approaches that of the full ambient permit market. Since the appeal of this approach diminishes over time, implementing it now would ignore the inevitable complexity just around the corner. The desirability of the single-market ambient permit system is purely transitory.

### Trading Rules

One final way to incorporate source location into permit-based pollution control policies involves the construction of trading rules. In one sense the trading rules approach is an amalgam of many of the best aspects of the other policies. Because it avoids zones, region-wide trades are permitted. The hot spot problem is eliminated without recourse to multiple markets by the selective use of transfer coefficients.

The trading rule approach represents a departure from the previously discussed approaches because it is not based on a complete set of markets, each with predetermined prices. Trading rules specify the procedures for deciding how much of an increase in emissions is allowed a purchaser or combination of purchasers given an emissions reduction by a seller or set of sellers. They focus on the transaction on a case-by-case basis rather than focusing on the market as a whole.[15] This approach presupposes that the attempt to define and market a separate concentration permit for each receptor is not likely to succeed. Yet it attempts to retain the use of transfer coefficients to govern trades among sources.

Three different trading rules have been proposed in the literature: (1) the pollution offset;[16] (2) the nondegradation offset;[17] and (3) the modified pollution offset.[18]

- The pollution offset approach allows offsetting trades among sources as long as they do not violate any ambient air quality standard.

- The nondegradation offset allows trades among sources as long as no ambient air quality standard is violated and total emissions do not increase. (Notice that this is, in effect, equivalent to the regulatory tiering approach in the Sulfur Allowance Program discussed earlier.)

- The modified pollution offset allows trades among sources as long as neither the pre-trade air quality nor the ambient standard (whichever is more stringent) is exceeded at any receptor. Total emissions are only indirectly controlled.

Consider the effect of these trading rules on two representative sources contemplating a trade (Figure 4-1). The trading possibilities are bounded by the various emissions and air quality constraints. The pre-trade situation is given as $(E_1, E_2)$, the currently allowed emissions rates for the two sources. $R_2$s and $R_1$s are, respectively, the emissions combinations between the two sources that allow the ambient air quality standard to be met at the second and the first receptors. As drawn, in the pre-trade situation only the second receptor is binding. R; defines the allowable emission combinations if current air quality (which, by construction, is cleaner than required by the standards) is to be preserved at the first receptor. The 45° line records the allowable emissions combinations that hold the level of total emissions constant at their pre-trade level.

In Figure 4-1, if the second source were to sell to the first source, the trading possibilities would be area A, if the modified pollution offset is used; A + B if the nondegradation offset is used; and A + B + C if the pollution offset rule is used.

Notice that if the second source is buying emissions reductions, all three trading rules yield the same set of trading possibilities (area D). The equivalence of the trading possibilities for all three rules in this context derives from the fact that current air quality at the receptor most affected by the purchasing source is already at the standard. Emissions would decrease and air quality would remain at the standard regardless of which of the three rules governed the trade.

Because costs are lowered as one moves away from the origin of this graph (since less control is required), after-trade allowable emissions would be rep-

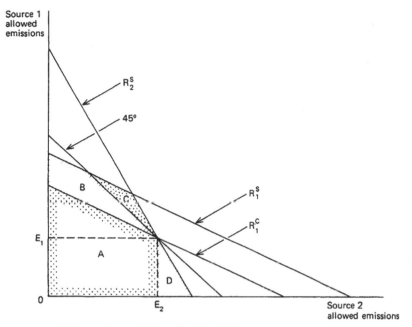

**FIGURE 4-1.** A Comparison of Three Trading Rules

resented by a point on the outer edge of the appropriate frontier. Those rules offering more trading possibilities normally would support lower cost allocations of the control responsibility. In this example, therefore, costs would be lowest for the pollution offset and highest for the modified pollution offset.

This potential cost advantage is one of two substantial advantages of the pollution offset. The second is that this is the only trading rule of the three that does not make the trading possibilities contingent on the pre-trade emissions situation for these two sources. This independence is clear from Figure 4-1. $R_1$s and $R_2$s are a function of the ambient standards and the amount emitted by sources other than the two being considered; they are not affected by the command-and-control allocation in any way. However, both the 45° line and $R_1$ are drawn given a particular starting allocation between these two sources. In the case of the U.S. Emissions Trading Program, this starting allocation is the command-and-control allocation. $R_1$ and the 45° line must pass through $(E_1, E_2)$.

This point is significant because it means that the degree to which the nondegradation offset and the modified pollution offset can be expected to diverge from the least-cost allocation is sensitive to this initial, administratively determined allocation. In essence, when the command-and control allocation defines the pre-trade equilibrium, these trading rules perpetuate excessive control. Since one of the clearest conclusions that comes from Chapter 3 is the large cost penalty associated with this excess emissions reduction, its importance to the non-degradation and modified pollution offset trading rules is an unfortunate characteristic. Although the cost penalty associated with these two trading rules would still be less, usually substantially, than the command-and-control pre-trade allocation, it never would be eliminated, except by coincidence.

Unfortunately, the attractiveness of the pollution-offset rule is marred by two other considerations: (1) any actual sequence of trades may fail to capture the potential cost savings; and (2) for some trades the ratio of emissions increases to emissions reductions would not be adequately defined by the rule.

Whereas Krupnick, Oates, and Van de Verg (1983, *24*) show that in the two-source, two-receptor case all gains from trade will be exhausted so that the pollution-offset trading equilibrium would be cost effective, it is not obvious that that result extends to the more realistic multiple-trade context. When a source reduces emissions, it triggers concentration reductions at a number of receptors. Any source increasing emissions will (by acquiring credits) similarly trigger concentration increases at a number of receptors. Even if these emissions increases and decreases were of the same magnitude, in general the concentration increases and decreases would not be of the same magnitude at every receptor.

In the pollution-offset system, depending on the location and stack heights of the trading sources, an emissions trade normally would create a vector of net concentration increases and reductions at various receptors. By design, only concentration increases triggering a violation of the ambient standard would have any bearing on the trade. Once the trade had been consummated, the trading parties would lose any responsibility for or claim over these concentration changes.

Compare this situation with an ambient permit system in which the seller has a well-defined property right over all of the concentration reductions resulting from its emissions reduction. Any selling source could treat concentration permits associated with different receptors as separate, tradable commodities, retaining those not specifically traded to the acquiring source. The retained credits could be sold to other sources as needed.

Whereas the ambient permit system retains the value of these reductions to the creating source, thereby encouraging their transfer to those sources valuing them the most, the pollution-offset system creates a "use it or lose it" situation. The property rights in the concentration permits not needed in the specific trade at hand presumably would be lost. Only those concentration permits relating to the receptor or receptors of interest to the acquiring source would receive full value. Because the rest could not be separated out and sold to other buyers, permits would not, in general, be transferred to those sources valuing them the most. Only if a series of transactions were consummated simultaneously and the valuable excess reductions were all acquired can this system result in a cost-effective allocation.

A second problem relates to defining the offset ratio, the amount of emissions reduction any expanding source has to acquire from another somewhat distant source in order to compensate adequately for each unit of expected emissions increase. According to the pollution-offset system, an acquiring source need only secure enough emissions reduction credits to prevent violations of the ambient standard. Suppose a source locates in a relatively clean portion of a nonattainment area where the most affected monitors are recording air quality levels that are quite a bit better than required by the standards. Suppose further that the increased emissions from this acquiring source would make the air quality worse at these monitors but would not trigger any violations of the ambient standard. In this case, according to the pollution-offset approach, the source would not have to gain any offsetting reductions from another source since the only constraint on the trade is the prohibition against ambient standard violations. In effect, sources affecting nonbinding monitors would be allowed to increase their emissions without securing any compensating reductions.

In essence, these two problems are related. In the first case, the problem results from inadequately defined property rights over created concentration reductions, preventing their transfer to those valuing them the most. In the second case, the problem results from the fact that no one holds property rights over the potential concentration units represented by the air quality that is currently better than the standards, allowing them to be allocated on a first-come, first-served basis. In both cases, because the prevailing property right structures create perverse incentives, the trading equilibrium normally would not be cost effective.

The two remaining types of trading rules attempt to provide some resolution of the second problem, but not the first. The nondegradation offset disallows any trades that increase emissions, while the modified pollution offset holds air quality at all specified monitors at least at current levels or at the ambient standard, whichever is more stringent. Both can be seen as crude ways of rationing the limited amount of "excess" air quality at nonbinding monitors by requiring some minimum amount of compensating offsets that every new or modified source must acquire.

Of these two means, the nondegradation offset is the simplest because the accounting system is much more straightforward. It only has to keep track of emissions and possible ambient standard violations. The modified pollution offset prevents trades that allow any concentration increases at any monitor.

Although actual cost savings from the pollution-offset and modified pollution-offset rules are not likely to coincide with the maximum possible cost savings for those systems, comparing the cost-effectiveness and emission-loading characteristics of these rules under circumstances that are the most favorable for their success still is instructive.

McGartland and Oates (1985, *15*) find that for particulate control in the Baltimore, Maryland, region, the modified pollution-offset system could achieve the pollution target at less than half the cost of the command-and-control approach, but it was still 70% more expensive than the pollution-offset approach. Interestingly, both systems resulted in more emissions than the command-and-control system. The excess emissions created by the modified pollution-offset trades were transported by local winds to the ocean and therefore did not degrade the air quality at local receptors. McGartland and Oates did not examine the nondegradation offset.

Atkinson and Tietenberg (1984) have examined all three systems in the context of particulate control in St. Louis, Missouri. The results indicate that when the primary ambient standard is the target, the nondegradation offset is only slightly more expensive than the ambient permit system. For two out of the three initial administrative allocations, the cost penalty associated with the use of the nondegradation offset was less than 10%. The modified pollution offset was not only more expensive, but it resulted in more emissions. Substantial savings were possible for all three trading rules compared with the pre-trade command-and-control allocations.

Since sulfur allowances are traded one-for-one and cannot be used in any location where their use would violate an ambient air quality standard, this combination of attributes is logically equivalent to the nondegradation offset rule. According to the results from the Atkinson and Tietenberg (1984) analysis, which predates the Sulfur Allowance Program, that combination may not have been a bad compromise.

## Summary

- Incorporating source location into an emissions trading program is a difficult but manageable proposition.

- Though theoretically able to meet the challenge, in practice ambient permit markets have not been used because of their inherent complexity. In addition, they are inconsistent with common statutory prohibitions against trades that allow emissions increases, since ambient permit markets usually allow more emissions than typical market approaches.

- Because they allocate excessive control responsibility to distant sources, unconstrained emissions permit trading systems typically result in much higher compliance costs than the least-cost solution. The cost penalty is particularly large when sources close to the most adversely affected receptors must control so much that they are on very steep portions of their marginal cost curves. This condition is relatively common in air pollution studies.

- Because they do not directly control concentrations, unconstrained emissions permit systems run a high risk of creating hot spots—pollutant concentrations that exceed the ambient standards at one or more points within the airshed. To provide a margin of safety for the ambient standards, authorities must compensate by making the cap more stringent, forcing even more over-control and yet higher costs.

- Zonal systems are one approach to controlling hot spots. Theoretical work suggests that the optimal number of zones will tend to increase with the degree to which the effects of pollution are localized. The more localized the effects of pollution, the greater the potential for hot spots, which will tend to favor a system with smaller trading zones. The slopes of the marginal benefit and marginal cost curves also matter, however, because they affect the cost savings from trading—steep marginal benefits favor smaller zones, while steep marginal costs favor larger zones.

- When the boundaries of zones are defined, pollution sources generally should be placed in zones with other sources that are close substitutes for them. Among other things, this important point requires consideration of the height of the discharge point in addition to its coordinates since a source with a tall stack can have different effects than an otherwise similar source.

- In practice, information is a key constraint on zone design. Full-information zonal permit systems afford more protection for ambient standards than emission permit systems and reduce compliance costs. However, control authorities would have to be omniscient if this approach were to work in practice; the information burden is unrealistically large. Limited-information zonal permit systems, those that can be initiated with reasonable amounts of information, are much less effective. Because zonal permits cannot be traded across boundaries, the cost penalty would be very sensitive to the initial allocation of zonal caps. No conventional rule of thumb for allocating this required emissions reduction among zones comes close to the cost-effective allocation.

- In practice, one alternative that has emerged is a system of directional trading in which one zone is allowed to acquire permits from another zone but not vice versa. For example, to prevent any shifting of pollution from the upwind zone to the downwind zone that might occur from trading, sources in the upwind zone would be prohibited from acquiring emissions permits from downwind sources, but downwind sources could acquire permits from upwind or downwind sources. The RECLAIM system illustrates this approach.

- One possibility for improving the performance of zonal permit systems is developing a procedure for allowing trades between zones based on an exchange rate (the ratio of the amount of emissions allowed in the buying zone for each unit of emissions reduced in the sellers zone). Theory makes it clear that it would not, in general, be possible to choose exchange rates such that the cost-effective solution would be reached. Empirically, however, rule-of-thumb exchange rates may lead to somewhat lower costs, as pointed out by the study that finds that adding exchange rate trading to the initial allocation of responsibility in controlling sulfur in Europe would reduce abatement costs on the order of 19%.

- Single-market ambient permit systems focusing on a single "worst case" receptor can typically come very close to the least-cost allocation with a minimum of transaction complexity. They also can afford a high degree of control over the hot spot problem given a stable configuration of sources and the absence of multiple binding receptors. Unfortunately, they also provide incentives for changing the spatial configuration of sources over the long run, undermining the selection of one particular receptor as the worst. Although adding new markets for every binding receptor would solve the problem, this expansion destroys the simplicity of the system, its most attractive feature. Since this approach sows the seeds of its own eventual destruction, its attractiveness is transitory.

- The establishment of trading rules that govern individual transactions, is, in principle, able to reduce cost penalties below levels associated with command-and-control levels, while affording greater protection from the hot spots problem.

- Of the three trading rules examined in the literature (the pollution-offset, the modified pollution-offset, and the nondegradation offset), the pollution offset is the closest to the ambient permit system. Although in theory it has the lowest potential cost penalty, in practice any likely sequence of trades would produce a trading equilibrium that is not cost-effective. If the trading is bilateral and sequential (as opposed to global and simultaneous), the outcome of the trading system is path dependent. Path dependence in this context means that the final equilibrium of the trading will depend on the trading sequence.

- The available empirical evidence suggests that actual cost savings from pollution-offset trades may fall short of cost savings estimated from models that assume global and simultaneous trading. However, the magnitude of the loss of cost savings due to bilateral trading can be reduced by allowing trades among individual sources rather than limiting them to large aggregates such as countries (an issue of some importance for greenhouse gas trading). The other disadvantage from pollution-offset trading is that if implemented, it would allow emissions to increase substantially beyond command-and-control levels.

- The modified pollution offset affords the most protection for current local air quality (although it does allow emissions increases from tall stacks). It shares with the pollution offset the weakness that unreasonably complicated trades would be required for the trading equilibrium to be cost-effective given the air quality constraints.

• One trading rule strategy, the nondegradation offset, has been implemented in the United States to prevent hot spots. Known locally as regulatory tiering, this approach applies more than one regulatory regime at a time. In the Sulfur Allowance Program, sulfur emissions are controlled both by the regulations designed to achieve local ambient air quality standards as well as by the sulfur allowance trading rules. All transactions have to satisfy both programs. Thus trading is not restricted by spatial considerations (national one-for-one trades are possible), but the use of acquired allowances is subject to local regulations protecting the ambient standards. Unlike hot spot prevention programs that either restrict all transactions or employ a much more strict cap, this approach prohibits only the few transactions that would result in a hot spot. Ex ante empirical analysis of this approach suggests that regulatory tiering may well have been an effective compromise in the Sulfur Allowance Program.

## Notes

1. Nitrogen dioxide is controlled both as a direct contributor to health problems and as a contributor to ozone formation. With respect to the health effects, it is a nonuniformly mixed assimilative pollutant, and therefore emission location matters.

2. See, for example, Ludwig, Javitz, and Valdes (1983).

3. The problems associated with the existence of market power are explored more fully in Chapter 7.

4. This kind of interdependency exists in controlling sulfur oxides and particulates; see Spofford (1984, 95–97).

5. "The term 'reasonable further progress' means annual incremental reductions in emissions of the applicable air pollutant ... which are sufficient in the judgment of the Administrator to provide for attainment of the applicable national air quality standards by the date required" (42 USC 7501).

6. It is, however, possible to meet ambient standards under the act by relocating emissions to attainment areas. Apparently, this has happened in the United States (Henderson 1996).

7. Neither the Hahn and Noll (1982) study nor the O'Ryan (1996) study provide a comparison of emissions reductions.

8. The rule-of-thumb states that given constant and common marginal control costs in each of two regions, total control costs for an emissions permit system used to achieve an ambient standard will decrease if the ratio of the original emissions permit

price to the price after regional expansion is greater than the ratio of the sum of the transfer coefficients for the near sources to that for the far sources.

9. This would be a least-cost allocation among the set of possible zonal allocations for a given set of zonal boundaries; it would not in general produce the regional least-cost allocation because the zonal permit system causes the marginal costs of emissions reduction to be equalized across sources within each zone, rather than the marginal costs of concentration reduction. The zonal permit allocation of control responsibility would coincide with the regional least-cost allocation in general only if each zone contained one and only one source and each such source received its cost-effective allocation.

10. These include McGartland (1984), Spofford (1984), and Atkinson and Tietenberg (1982).

11. Spofford (1984, *82*) and Atkinson and Tietenberg (1982, *120*).

12. Atkinson and Tietenberg (1982, *119*) and Spofford (1984, *81*).

13. Harrison (2002) reports that the original RECLAIM proposal envisioned 38 separate regions, but this was abandoned due to fears that the markets would contain too few participants.

14. As the author points out, this result exactly parallels the classic finding of Weitzman (1974) that if the marginal-cost curve is steeper than the marginal-benefit curve, the regulator should pay more attention to the cost of control than the amount of control (i.e., emissions reduction) and should, therefore, use the price instrument.

15. Notice that it is this individual transaction focus that differentiates trading rules from the zonal exchange rates discussed above.

16. Krupnick, Oates, and Van de Verg (1983).

17. Atkinson and Tietenberg (1982).

18. McGartland and Oates (1985).

# 5

# The Temporal Dimension

The temporal pattern of emissions rates and whether the resulting concentrations accumulate or dissipate can be important factors in the design of any air pollution control strategy, including emissions trading. This chapter examines how emissions trading programs handle the temporal dimension.

Three specific aspects of the temporal dimension are considered. First, how much temporal flexibility is (and should be) allowed when permits are used, via banking and borrowing? While flexibility clearly can lower compliance cost, it also can admit the possibility of a temporal accumulation of concentrations. How can this potential conflict of goals be reconciled? Second, to what extent can (and should) emissions trading designs accommodate seasonal considerations for those pollutants, such as ozone, that exhibit seasonality? And finally, this chapter will examine how emissions trading can deal and has dealt with "episodes"—temporary, but potentially damaging, periods of especially high concentrations.

This chapter will examine how cost-effective or cost-efficient strategies can be developed in light of these considerations and the extent to which current treatments of the temporal component of the current policy can be improved.

## Borrowing, Banking, and the Nature of the Environmental Target

The design of the temporal component depends on the nature of the environmental goal. If the goal is defined purely in terms of emissions reductions, it is possible to allow considerable temporal flexibility without posing an environmental risk. If, however, the goal is defined in terms of ambient concentrations or environmental damage, shifts in emissions from one time period to another could lead to a clustering of emissions or affect the rate of accumulation. Higher concentrations (hot spots), due either to emissions clustering or accumulation, can cause more damage.

The potential for hot spots affects how the temporal fungibility of permits is treated. In cap-and-trade systems, the allocated permits usually are dated with the year of allocation. In systems without temporal flexibility, the permit can only be used during that year. For this design, the environmental concern arises over the timing of the permit use within that year, not over the possibility of shifting emissions to another year.

Emissions trading systems with temporal flexibility can incorporate either banking or borrowing or both. Banking means holding a permit beyond its designated year for later use. Borrowing means using a permit before its designated date.

The economic case for banking and borrowing is based upon the additional flexibility it allows sources in the timing of their abatement investments. Flexibility in timing is important not only for reasons that are unique to each firm but also for reasons that relate to the market as a whole. The optimal timing for installing new abatement equipment or changing the production process to reduce emissions can vary widely across firms. This can reflect such factors as the age of the current equipment that is being replaced, as well as the moving target of technological options for additional control that each firm faces.

Price considerations also argue for temporal flexibility. If all firms were forced to adopt new technologies at exactly the same time, the concentration of demand at a single point in time (as opposed to spreading it out) would raise prices for the equipment as well as for the other resources, such as skilled labor, necessary for its installation.

Banking also has the potential to reduce price instability. Storing permits for unanticipated outcomes, such as an unexpectedly high production level that triggers higher-than-expected emissions, can reduce future uncertainty considerably. Because stored permits can be used to achieve compliance during tight times, they provide a safety margin against unexpected contingencies.

While the economic case for allowing flexibility is compelling, the environmental case is more mixed. Allowing this kind of temporal flexibility can either ameliorate or exacerbate pollutant concentrations. If firms use the flexibility to disperse emissions through time, concentrations will be diminished. However, if this flexibility results in clustered emissions, concentrations will be worsened.

## Linking Emissions and Pollutant Concentrations

This section begins by examining what we know about the physical links between the timing of emissions and the resulting concentrations.

## Sources of Concentration Variation

Pollutant concentrations at specific receptor locations are monitored by taking samples of the air at frequent, regular intervals. Typical plots of concentration frequencies recorded from those samples show the ambient concentration to be distributed as a log-normal random variable.[1] The log-normal distribution is nonsymmetric, with the modal (most frequent) concentration lower than the mean (average).

Variation in emissions rates is a dominant source of the variation in concentrations. Even in the absence of emissions trading, sources do not emit at constant rates. Some emissions rates show a striking seasonal or daily pattern. Space heating or air conditioning emissions are seasonal, while mobile source emissions increase significantly during morning and evening rush hours. Others caused by breakdowns or accidents may be more random. With emissions trading that allows banking or borrowing, firms can vary their emissions rates over time even more.

Variation in meteorological conditions is a second source of concentration variation. Wind speed and direction both have random and cyclical components, including a definite seasonal pattern, particularly for areas near large bodies of water. Since the water and the land heat and cool at different rates, local onshore or offshore breezes can be created by this temperature differential.

Thermal inversions are the most adverse weather conditions from the point of view of pollution control. They occur when a temperature inversion distorts the normally smooth upward flow of warm air. Thermal inversions rob the atmosphere of its normal ability to disperse and dilute the pollutants, trapping emissions in a small volume of air and creating very high concentration levels.

A final source of concentration can occur when the emitted substances accumulate over time rather than dissipate. Accumulation can occur whenever the amount of substance added by emission exceeds the amount that is depleted from the atmosphere by chemical reactions or settling out.

Concentration variations have important consequences for policy. In the United States, for example, the Clean Air Act bases the primary ambient standards on human health. Are short-term, high concentration exposures harmful, or is human health more sensitive to cumulative exposure over longer periods of time? How should policy be structured to reduce short-term or long-term exposure?

## The Role of Ambient Standards

For each pollutant, the form of the U.S. ambient standard has been chosen to mitigate the known or suspected damage caused by that particular pollutant. If short-term, high-dose exposures are found to be dangerous, the standard is stated in terms of maximum, short-term exposure. Short-term exposures are monitored using concentration levels averaged over periods as short as an hour or as long as 24 hours. If the damage seems to be closely related to longer-term exposure, then the standard is stated in terms of an annual average.

Several pollutants have multiple ambient standards.[2] Sulfur oxides and particulates have an annual average standard and a 24-hour standard. Nitrogen dioxide has only an annual standard. Carbon monoxide has an 8-hour standard and a 1-hour standard. When multiple standards are established, the states must comply with all of them.

The form of the standard makes a difference in defining the strategies to meet it. Ambient standards based purely on annual averages can be met without worrying about the timing of the emissions within the year. Emissions reductions reduce the annual average whenever they occur. The fact that concentration levels exhibit a series of peaks and troughs over time is of little consequence for annual average standards.

When the ambient standard is based on a short-term average, however, both the timing and the quantity of emissions are important. Since the objective is to reduce the highest concentrations, one strategy is to shift emissions away from the peak concentration periods. Although this strategy would have a large impact on reducing the highest concentrations, it would not necessarily have any impact on the annual average. Indeed, it is conceivable that strategies that are successful in meeting the maximum short-term average standard could leave the annual average concentration unaffected or could even increase it, providing that off-peak emissions increased by a large enough amount.[3]

Temporal strategies that control the timing as well as the level of emissions become more attractive as the averaging period is reduced. Since shorter averaging times imply less opportunity to reduce the average by combining peak with off-peak periods, reducing emissions during a relatively short period is especially important. To the extent that these standards are picking up concentration spikes—short-duration, high-concentration peaks—shifting emissions to other periods becomes easier, since the change in emissions timing involved in any shift of emissions from a peak to an off-peak concentration period can be rather small.

When the major component of the variation in observed concentration is regular and, therefore, predictable, strategies that seek to reduce peak concentrations are feasible. Peak concentrations for certain pollutants always occur during the same season or even the same time of day, allowing their occurrence to be anticipated and controlled. In these cases, peak concentrations can be associated with specific times.

The point is illustrated most easily by tropospheric ozone, the pollutant most frequently responsible for regions receiving a nonattainment designation in the United States. The ambient standard for ozone is based on both an 8-hour averaging time and a 1-hour averaging time. One-hour ozone concentrations typically show pronounced diurnal (daily) and seasonal peaks. In the nonattainment areas in the northeastern United States, for example, the standard is exceeded only during the summer. The prevalence of ideal conditions for ozone formation—warm sunny days—during that period is responsible. These pronounced daily and seasonal patterns, coupled with a standard defined in terms of a short averaging time, make ozone a prime candidate for controlling the timing, as well as the level, of emissions. As noted in Chapter 1, two emissions trading programs have capitalized on that opportunity.

# The Role of Banking and Borrowing

## *The Cumulative Emissions Target Case*

The easiest case for emissions trading design is where only the cumulative level of emissions matters. In this case, the timing of the emission is not important. The unimportance of emissions timing has several immediate design implications. Permit dating is not important. With undated permits, the issue of banking and borrowing becomes moot; in effect, this design automatically allows unlimited banking and borrowing.

As was pointed out in Chapter 2 (and developed earlier in Tietenberg 1985), in this system the price of permits would normally rise at the rate of interest and the holders would automatically choose to use them in the manner that minimizes the present value of abatement costs. The incentives created by decentralized decisionmaking in this case are compatible with social objectives.[4]

Rubin (1996) extends this analysis by providing a more general treatment of permit trading in continuous time through the use of optimal-control theory. Instead of limiting intertemporal trading to banking, his paper considers the effects of allowing both borrowing and banking. His results show that since future abatement costs are discounted, the firm's emission streams will normally decline through time.

## *Banking and Borrowing When Emissions Timing Matters*

How should emissions trading deal with those cases where emissions timing also matters? Examining this question requires the use of an efficiency, rather than a cost-effectiveness, framework. Efficiency models consider not only the damages caused by pollution but also how emissions timing affects the nature of those damages.

Kling and Rubin (1997) point out that this situation opens the door to a potentially important market failure. Firms have an incentive to minimize the present value of abatement costs but not the present value of all costs, including the damage caused by emissions. In general, the resulting (inefficient) incentive is to delay abatement—abating too little during the early periods and concentrating too much abatement later. This concentration of emissions in the earlier periods raises the present value of total costs, including both abatement costs and damages, above the efficient level.

Under their base case, which involves temporally stable costs, output prices, and emission targets, the firm's incentive to borrow emissions from future periods (due to the effect of discounting) exacerbates the clustering of emissions in the present. When firms use a higher discount rate, the emissions stream declines even more quickly, implying increased borrowing.

In other circumstances, such as when marginal abatement costs rise, marginal production costs fall, emission targets become more stringent, or output prices rise, Kling and Rubin (1997) find that firms may have an incentive to bank, rather than borrow, permits. In these cases, two opposing forces affect firms' timing decisions: (1) higher future marginal abatement costs, lower future marginal production costs, or more stringent future emissions targets encourage firms to reduce more in early periods when costs are relatively low; but (2) a positive discount rate encourages firms to delay cleanup to later periods. If the effect of increased future needs for permits is greater than the discount rate effect, firms have an incentive to bank permits.

## *Solutions*

Kling and Rubin (1997) suggest that in those cases where the borrowing incentive creates an unacceptable degree of clustering, restorative design changes should be considered. Instead of one-to-one trading between time periods, their analysis proposes modified banking rules that would require firms to discount borrowed emissions. (Note the similarity of this proposal to the use of exchange rates for spatial trading discussed in Chapter 4.) Like a more typical financial bank, these modified rules would require sources to pay interest on borrowed permits.

As those authors point out, interestingly, to the extent that regulators have modified the intertemporal trading ratio, they have done so in the opposite direction. For example, the California Low-Emission Vehicle Program discounts banked (not borrowed) emissions by 50%, 75%, and 100%, 2-, 3-, and 4-model years after the credits were earned, respectively. This banking program thus discourages saving permits in early periods for use later rather than rewarding the banking of permits. As a result, it encourages the clustering of emissions in the present.

Another example is provided by the $NO_x$ Budget Trading Program (NBP), which incorporates a "progressive flow control" banking system. The NBP's progressive flow control provisions were designed to discourage concentrating the use of banked allowances in a particular future ozone season. Flow control is triggered when the total number of allowances banked for all sources exceeds

10% of the total emissions budget for the following year. When flow control is triggered, the EPA calculates the flow control ratio by dividing 10% of the total budget by the number of banked allowances, implying that a larger bank will result in a smaller flow control ratio. The resulting flow control ratio indicates the percentage of banked allowances that can be deducted from a source's account in a ratio of one allowance per ton of emissions. The remaining percentage of banked allowances, if used, must be discounted and deducted at a rate of two allowances per ton of emissions. This approach also discourages banking whenever flow control is likely to be binding.

What about prohibiting intertemporal trading by eliminating both banking and borrowing? Is it possible to evaluate the costs and benefits of trading to come up with some rule-of-thumb prescriptive findings?

It turns out that it is, even in the face of uncertainty about how trading would affect the temporal distribution of emissions. Yates and Cronshaw (2001) show that from an efficiency point of view, the regulator should allow intertemporal trading if and only if the slope of the marginal damage function is less that the slope of the marginal aggregate abatement cost function.[5]

This theoretical result has an important implication. As more pollution is controlled, the marginal cost of abatement is expected to rise and the marginal damage caused by remaining emissions is expected to fall. Combining this empirical expectation with the Yates and Cronshaw result implies that the desirability of intertemporal trading will become more important as the stringency of control rises.

Of course, not all programs involve this level of uncertainty about the emissions distribution. Schennach (2000) analyzes the economics of allowance banking in the context of the Sulfur Allowance Program. Two characteristics of that program are important in her analysis. First, the program allows banking but no borrowing, and second, the emissions target in Phase II is significantly more stringent than the emissions target in Phase I. She shows that the evolution of the firms' behavior over time can be separated into two periods: a banking period where units save part of their annual permit allocation for future use, followed by a period where all permits allocated annually are used immediately and the stock of banked permits is drawn down. In this case, intertemporal trading encourages early reductions, which produces a positive environmental benefit.

Schennach (2000) also notes that sources face considerable uncertainty about the future as they plan abatement investments. What effect does this uncertainty have on the incentive to bank emissions? The fact that the market as a whole cannot borrow allowances from future allocations introduces another incentive to bank in the face of uncertainty. An unexpected reduction in electricity demand could be accommodated by saving more allowances than expected during that period, thereby smoothing the effect of this shock on the price level. An unexpected increase of the same size, however, only could be absorbed by using allowances from the existing bank. This buffering capability is limited, since an optimal smoothing could require more allowances than available at that time. Hence, when depleting the allowance bank at some future time is a possibility, sources plan to bank more allowances than would otherwise be the case.

Actually, the conclusion that uncertainty promotes banking has not gone uncontested. Ben-David et al. (2000) investigate the effects of uncertainty and concomitant risk aversion as they affect the incentives of an emissions permit market modeled after the U.S. $SO_2$ market. In an experimental laboratory setting, they find that the irreversibility of investment in abatement technology that could subsequently prove to be excessive could present countervailing incentives for potential sellers of permits, resulting in a wait-and-see attitude toward adoption of efficient levels of abatement technology. In the actual Sulfur Allowance Program, a considerable amount of banking took place.

Borrowing is prohibited in the Sulfur Allowance Program. What was the effect of this prohibition on cost savings? Ellerman et al. (2000, *282*) calculate that of the $20 billion savings that can be attributed to emissions trading in the Sulfur Allowance Program, $1.3 billion was due to banking. Though in the presence of positive discount rates borrowing can lower the present value of abatement costs, its elimination does not sufficiently undermine the economic case for intertemporal trading as to render it ineffective, particularly in the case of declining emissions targets.

Subsequent analyses (Ellerman and Montero 2005) that evaluated the efficiency of observed temporal patterns of abatement based on aggregate data from the first eight years of the Sulfur Allowance Program, which involved a considerable degree of banking, find the resulting pattern of emissions to be reasonably efficient. This is an especially interesting finding since some had speculated that the high degree of banking during the first phase was simply an overreaction to the uncertainty created by a new program with stringent goals.

## *Price Stability*

Does banking increase price stability? That turns out to be a difficult question to answer. Ex post evaluations have a difficult time ferreting out the specific role of banking. In the Sulfur Allowance Program, for example, prices clearly have stabilized, but it is not clear what the driving forces have been. The transparency of prices achieved through the Chicago Board of Trade auctions certainly played some role by making the magnitude of market clearing prices public information, but so did banking.

We do, however, have the benefit of laboratory experiments. Godby et al. (1999) conducted an experiment to investigate how banking affects prices under uncertainty. In this particular experiment, information overload for the subjects was reduced by computerized advice on the optimal allocation of permits across periods and their implied marginal values. In this experiment, while uncertainty about the control of emissions led to substantial price instability when banking was not allowed, banking virtually eliminated the instability.

### *Operating Program Choices about Banking and Borrowing*

Tradable permit schemes differ considerably in how they treat banking or the role of forward markets.[6] No existing system that the author is aware of is fully temporally fungible, meaning it allows complete banking and borrowing.

Programs differ considerably in their approach to banking but not borrowing. The ETP allowed banking but not borrowing. The Lead Phase-out Program originally allowed neither, but part way through the program it allowed banking. The Sulfur Allowance Program has banking but not borrowing, and the RECLAIM program has an overlapping timeframe for compliance that is equivalent to a highly restricted banking and borrowing system. Banking of excess reductions for future years is allowed within the first phase of the EU ETS, but banking between phases is at the discretion of member states between the 2005–2007 start-up phase and the 2008–2012 commitment period. The Chicago ERMS rule provides that tradable units have a limited, two-year lifetime.

How important is temporal flexibility? The message that emerges from this review is that this temporal flexibility afforded by banking can be quite important:

- As noted above, Ellerman et al. (2000, *282*) calculate that about $1.3 billion of the $20 billion cost savings in the Sulfur Allowance Program could be attributed to banking and that the program promoted early reductions that led to less temporal clustering of emissions.

- Harrison (2004) reports that during the tremendous pressure placed on the market by the electricity generation problems in California, even the limited temporal flexibility in RECLAIM[7] allowed the excess emissions to be reduced by more than a factor of three—from about 19% to 6%.

- Nussbaum (1992) reports that the three years of banking in the U.S. Lead Phase-out Program not only reduced costs by about $200 million but produced much earlier reductions in lead than otherwise would have been achieved.

- Pendersen (2003) also suggests that temporal flexibility has been important for investment in the Danish greenhouse gas program.

## Strategies for Controlling Seasonal or Episodic Peaks

Controlling the timing as well as the flow of goods and services is a familiar activity in public policy. Peak-hour pricing is common among public service providers, not only as a better way of utilizing the existing generating capacity, but also as a way of lowering the demand for new capacity. Charging peak-hour use at a higher rate simultaneously forces those who create the need for new capacity to bear the cost of building it and to consider switching some of their peak use to off-peak periods as a viable alternative to new construction.

Telephone service and mass transit pricing provide two more familiar examples of peak-hour pricing. Higher rates are charged for calls placed during the peak periods, both to finance the capacity expansion needed to meet the growth in calls during the peak period and to encourage off-peak use when possible. Higher peak-hour transit fares, common in cities such as Washington, DC, are designed to accomplish a similar balance between peak and off-peak transit travel.

Although these familiar examples make it clear that temporal control is not a particularly novel concept, they all involve charging higher prices in peak periods. Since the emissions trading program is a quantity-based approach, not a price-based approach, the control authority regulates emissions, not prices. How could cost-effective temporal control be exercised in a quantity-based system?

In thinking about incorporating the timing of emissions into a transferable permit approach, it is important to distinguish two different types of temporal control. The first involves regular, and therefore anticipated, seasonal or daily fluctuations in concentrations. This type of control will be referred to as periodic control to convey the regularity of these conditions. The key aspect of periodic control is that the timing of these conditions can be anticipated. Because the conditions can be identified in advance, the means of controlling emissions during those periods also can be identified in advance.

The second form of control, known as episode control, involves irregular concentration peaks, which, to the extent they can be anticipated at all, can be anticipated only very shortly before they occur. The timing of these peaks depends on the initiation of thermal inversions or other adverse meteorological circumstances that occur randomly. The necessity for additional control cannot be identified more than a day or so in advance.

## Seasonal Control

Perhaps the easiest way to describe control that is targeted to specific periods is to compare it with constant control. Constant control defines the control responsibility in terms of a temporally invariant allowable concentration level and a temporally invariant mix of source responsibilities for meeting that level. In contrast, both the allowable concentration level and the mix of source responsibilities vary over time with targeted temporal control.

The recorded pollution at any monitor is composed of two elements: background and controllable pollution. Background pollution results from sources not under the control of the local control authority, either because they are unregulated or because they are located outside of the control jurisdiction. Controlled sources must make up the difference between the observed reading and the ambient standard. As the background pollution levels change, such as when ozone transported from other regions increases, the remaining allowable concentration changes as well.

The mix of source controls changes over time for a variety of reasons. Costs of control for each source vary across seasons or even over the 24-hour daily cycle. For non-uniformly mixed pollutants, the transfer coefficients may change as a result of shifts in the prevailing meteorology, or different monitors may be recording exceedances in different seasons, necessitating changes in both the degree and location of emissions control over time.

## Uniformly Mixed Pollutants

How would a temporally targeted permit system take these considerations into account? For uniformly mixed pollutants, location is not important, so transfer coefficients would play no role. The first step in controlling these pollutants would be to designate the number of permit periods.[8] This could be as simple as two periods corresponding to a peak concentration and an off-peak concentration period or as complicated as necessary. Using the chemical reactivities, meteorological conditions, and background pollution levels unique to each period, allowable emissions would be scaled to meet the standard, with individual permits designed to conform to that allowable emissions level.

Consider how this system applies to the control of ozone. Because ozone formation depends on temperature and sunlight, it is not a problem in the North during the winter months. The warm months could be designated as the peak season and the cold months as the off-peak season. The specific temporal boundaries between the peak and off-peak periods could be defined in terms of temperature or sunlight and could vary from region to region. More baseline control responsibility would be assigned in the peak season. Trades reducing emissions during the peak season would be especially encouraged.

While historically emissions trading programs took no account of seasonal factors in controlling pollutants, the NBP represented a significant change. Under the NBP, EPA established an ozone season $NO_x$ budget for each affected state. Recognizing the importance of seasonality in ozone formation, this budget caps emissions from May 1 to September 30. This approach also is used in the Chicago VOM emissions trading system.

## Non-Uniformly Mixed Pollutants

The cost-effective permit system would be more complicated for non-uniformly mixed assimilative pollutants. The environmental target would be defined in terms of allowed concentrations rather than emissions. The number of permits allocated to each period would be derived from, and defined in terms of, allowed concentrations in each period. Whereas uniformly mixed pollutants could be traded on a ton-for-ton basis within each defined period, trades among sources involving non-uniformly mixed assimilative pollutants would have to use the transfer coefficients pertaining to the particular period of the trade. Trades involving concentration reductions during the peak season would have to use transfer coefficients computed especially for that period.

The essence of this system would be to tailor the degree of control to the need for it. Sources would be confronted with the need to control emissions especially vigorously during those periods where the most control was needed to meet the standards. Because of their relative scarcity, the permits for the peak period would be priced higher. Those higher prices would provide sources with an incentive to add extra control during those periods, to switch some of their emissions to off-peak periods, or to control more for all periods, selling some of the surplus off-peak control.

## The Constant Control Alternative

Constant control ignores the greater damage done by emissions during a peak season and designs emissions standards with sufficient stringency that concentrations fall within the standards even under the most adverse meteorological circumstances. A constant control strategy would shift the entire distribution of concentrations toward the origin, whereas a periodic control strategy would reduce the right-hand tail of the distribution.

Constant control is an excessively costly means of reaching a short-term standard for two reasons: (1) it requires larger emissions reductions than necessary to meet the ambient standards; and (2) it does not take into account which sources can control most cheaply at the time the control is needed. Larger emissions reductions are required by a constant control strategy because all sources are required to undertake a degree of control that is sufficiently stringent to meet the ambient standard under the most adverse conditions. This "worst case" approach requires more control than necessary in less adverse circumstances.

Because it allocates the control responsibility among sources so as to minimize the cost of constant control rather than the cost of periodic control, a constant control strategy fails to identify and take advantage of sources that can exercise control most cheaply during those periods when the largest amount of control is needed. Some sources may well be easily induced to exercise more control during the peak season with a higher peak permit price. Without allowing the permit price to reflect the time variation in the difficulty of control (as would be the case with constant control), sources that could control most easily during that period never would be identified. As a result, they control too little during that period, while other sources control too much.

The cost of a constant control strategy seems to be sensitive to the stringency of the short-term standard. The more stringent the standard, the greater the cost of constant control. Investigating the sensitivity of the cost of controlling nitrogen dioxide in Chicago, Anderson, Reid, and Seskin (1979, 5-28 to 5-34) find that using a constant control strategy, a 500 g/m$^3$ 1-hour standard could be attained at an annual cost of $1 million, while $24 million would be required to reach a 250 g/m$^3$ 1-hour standard. Though they did not analyze how much this cost would be reduced if periodic instead of constant control were used, they did find that the cost of constant control rises very rapidly as the standard is made more stringent. Because meeting more stringent standards with constant controls creates larger excess control capacities during the off-peak periods, the potential for large savings from periodic controls is enhanced as long as flexible technologies are available.

## Cost Savings

Though the literature provides few air pollution studies that shed light on the magnitude of cost savings due to the use of periodic rather than constant control, fortunately there have been some relevant water pollution studies. Though the circumstances are sufficiently different in air and water pollution control to preclude using the results from one to draw firm conclusions about the other and the studies are rather dated, the results are suggestive and included for that reason.

One case study by Yaron (1979) involved two reaches of the DuPage River in Illinois that were being polluted by two treatment plants and four industrial plants. The model contained two seasons and simulated the achievement of a dissolved oxygen standard by both constant and periodic control. The results suggest that periodic control is significantly cheaper. The total variable cost was $400,405 per year for the constant control, but $137,046 less for the periodic control—a savings of 34%.

A second study by O'Neil (1983) is particularly helpful because of its detailed treatment of capital costs and source location in a multiple receptor framework. Simulating biochemical oxygen demand emissions into the Fox River in northern Wisconsin for each of three periods, he compared a periodic policy with a worst case, temporally invariant rule. The Wisconsin Department of Natural Resources defined the worst case adverse flow and temperature conditions, and a constant matrix of transfer coefficients derived from these conditions was used to govern trades in all three periods. The periodic case involved different transfer coefficients in each of the three periods.

Two principal results were obtained. First, the worst case rule based on flow and temperature conditions failed to prevent the standard from being violated at one of the receptors during one of the off-peak periods. This result underlines an additional problem with constant control—it is difficult to select in advance a single transfer coefficient matrix and a single number of permits that will achieve the standards in all periods.[9] By using transfer coefficients and allowable emissions tailored to each period, the periodic policy reduces this problem.

The actual concentration in any period is a function of the background concentration plus a weighted average of the emissions from all sources, where the weights are the prevailing transfer coefficients. A constant policy uses only one set of transfer coefficients for the entire year, which would be accurate in only one period (the designated worst-case period). Therefore, the use of a single constant matrix opens the distinct possibility that the standards could be violated in the other periods. That is precisely what happened in the case study by O'Neil. Constant policies increase the risk of violating the ambient standards.

Although for most situations there would exist a true worst case that could eliminate the problem, in general the control authority cannot define that worst case in advance. It depends not only on the allowable concentration increases and the transfer coefficients (both of which can be determined in advance) but also on the distribution of emissions loadings among sources for each period (which cannot be determined in advance). Therefore, the control authority has only two undesirable choices with a constant policy: It can either include a significant margin of safety (extra control), which is expensive, or it can run the risk of violating the ambient standards, which over the long run could be even more expensive.

The O'Neil study also finds that the constant worst-case scenario resulted in control costs that were approximately the same as those for the periodic policy. This result is due to the type of abatement modeled in this study in which variable costs were quite small in comparison to fixed costs. Since only variable costs can be saved, the dominance of fixed costs leads to small estimated savings. Though their costs are not strictly comparable because the periodic policy met the standards while the constant policy did not, this result does point up the importance of the balance between fixed and variable costs.

In practice, existing firms with installed equipment would find the cost savings from temporal control more difficult to capture than would new sources. Even with a periodic control policy, the capacity of any capital control equipment would have to be tailored to the period representing the most severe emissions reduction requirements for that source, leaving excess capacity in the other periods. Because this capacity has to be financed whether or not it is fully used, nothing is saved when installed capacity is underutilized. In contrast, operating costs are typically variable and, therefore, can be saved in periods of lower control. Because sources with high variable and low fixed control costs would be able to take more advantage of a periodic policy, capturing more cost savings, the ability to save costs with periodic policies depends on the mix of control technologies. Periodic policies encourage the development of more flexible technologies and accommodate their flexibility.

One possible concern with periodic controls is that their use could lead to larger emission loadings on the environment as emissions from peak concentration periods are shifted to other periods. Can a workable plan be offered in the face of this objection?

The emissions loading issue can be dealt with through the judicious use of long- and short-term ambient standards. All emissions enter into the calculation of the annual average, so it provides a convenient check on emission loadings. As long as the ambient standard is defined in terms of a short-term and an annual average, the annual average would control emission loadings while the short-term average would be used to ensure that health is protected from high, short-term exposure. For those pollutants having standards defined only in terms of a short term standard, an annual average could be added as needed. For those pollutants having annual averages, the levels could be made more stringent to reflect their new purpose, if that is necessary.

Care would have to be taken for pollutants that can be transported long distances to ensure that emissions from tall stacks would not escape this accounting device. Otherwise, unwanted higher emission loadings could occur with periodic approaches, just as they could with spatial approaches.

Unreasonably high control costs also open the door to variances. Faced with crippling costs, sources seek to reduce their burden through the courts. Their chances for success are enhanced when the costs seem totally out of proportion with the benefits. Approved variances result in higher emission loadings. Because periodic control costs less, the demand for variances and the chances of variances being approved are diminished. While in principle emission loadings could be higher with a periodic policy, in practice they could be lower as well.

## Episode Control

To the extent that thermal inversions or other devastating meteorological circumstances occur regularly at the same times during the year, in principle they could be handled by a periodic policy. In practice, that is not possible because the occurrence of these conditions is not regular and the degree of control required is so severe. This high degree of control should be imposed only when needed despite the fact that the periods of need cannot be identified more than a day or so in advance. Episodes deserve separate procedures and market designs.

To be cost-effective, episode control policies need to identify in advance those sources that can cut back relatively cheaply on very short notice. Meteorologists can give at least one-day advance notice and sometimes more as to when the episodes could occur. The basic problem with identifying the amount each source should reduce emissions during an episode is that every source has an incentive to argue for the smallest control responsibility possible. Because they bear the cost of further control, they can hold costs down by avoiding responsibility.

An episode permit system offers one way to resolve this problem (Howe and Lee 1983). It could be a relatively simple program whenever location could safely be ignored. Unless the episodes at some particular location tend to be triggered by one particular source, the primary object is to reduce total emission loadings in a region. This is the objective most likely to support an episode permit system.

An episode permit system would be initiated by assigning priority numbers to emission permits. These either can be assigned continuously for each ton of emissions or, as is probably more practical, assigned to a few priority categories. The higher the priority of the right, the lower the probability that its use would be prohibited during an episode. The highest priority permits would allow uninterrupted, continuous emissions.

Consider an example with two priority categories. During a priority-one period (when emissions would be the most restricted), only priority-one permits could be used. During a priority-two period, either priority-one or priority-two permits could be used. The control authority would be responsible for defining the conditions that would trigger a priority-one or a priority-two alert, as well as announcing via a prearranged communication channel when these alerts were in effect. It also would be responsible for defining the number of permits (i.e., the amount of allowed emission) during each type of alert and enforcing the ultimate allocation.

Once the system had been established, sources would secure the appropriate number of permits for each priority designation by trading among themselves. Since priority-one alert designation permits would allow uninterrupted, continuous emissions, they would command the highest price. Many sources could be expected to purchase some of each category of permits, reflecting their costs of short-notice reaction. They would cut back emissions during each type of alert until the marginal cost of control was equal to the permit price for that priority category.

With this system in place, the control authority would announce the alert number and all sources would reduce emissions to the predetermined levels established for that alert. Because of the different prices of permits, sources would voluntarily sort themselves out by their costs of short-notice control. Those who could control relatively cheaply on short notice (say by switching fuel or using afterburners for volatile organic compounds) would save money by not having to purchase the priority-one permits. Others who could not reduce emissions as cheaply would purchase the permits and continue emitting during the episode.

The episode permit system not only provides incentives for existing sources to make short-notice reductions as cheaply as possible, it also provides incentives to develop and to adopt control technologies that allow this kind of flexible response. By adopting more flexible control systems, sources could lower their expenditure on episode permits. By allowing sources to save money as they adopt flexible-control technologies, control authorities give the manufacturers of these technologies an edge over competitors who are producing less flexible control technologies. The cost savings from an episode emissions trading system could be expected to rise over time.

Only one published, unfortunately quite dated, empirical study attempts to establish the magnitude of savings possible from using an episode control policy tailored to the circumstances rather than a constant control policy that is sufficiently stringent to preclude the episodes under any circumstances. Teller (1970) examined the cost of controlling sulfur dioxide in Nashville, Tennessee, using fuel substitution for constant and episode control strategies. Although both strategies were designed to ensure attainment of the same ambient standard, the constant control strategy was found to be five times more expensive than the episode control strategy.

# Summary

## *Banking and Borrowing*

- If the environmental target can be satisfied solely in terms of emissions reductions, it is possible to allow considerable temporal flexibility without posing an environmental risk. If, however, the goal is defined in terms of ambient pollutant concentrations or pollutant damage, shifts in emissions from one time period to another could lead to a clustering of emissions. Temporally concentrated emissions lead to higher peak concentrations (hot spots) than those that are more temporally dispersed.

- Emissions trading systems can incorporate temporal flexibility by allowing banking, borrowing, or both. Banking means holding a permit beyond its designated date for later use. Borrowing means using a permit before its designated date.

- The economic case for banking and borrowing is based upon their allowing sources considerable flexibility in the timing of their abatement investments. Flexibility in timing is important not only for reasons that are unique to each firm, such as the age of existing equipment, but also for reasons that relate to the market as a whole, such as the need to avoid unnecessary price increases as all firms seek to acquire the same new equipment at the same time.

- When only the cumulative level of emissions matters, the price of permits normally would rise at the rate of interest and the holders automatically would choose to use them in the manner that minimizes the present value of abatement costs. Decentralized decisionmaking is compatible with social objectives for this case.

• When a cumulative emissions cap is not sufficient to protect against damage from concentration peaks, timing must be considered as well. Situations where the damaging effects of peak concentrations are important open the door to a potentially important market failure. Firms have an incentive to minimize the present value of abatement costs but not the present value of all costs, including the damage caused by emissions. In general, the resulting incentive is to delay abatement (abating too little during the early periods and concentrating too much abatement later).

• Delaying abatement is not always the optimal choice for a firm, even in an unrestricted permit market. When marginal abatement costs rise, marginal production costs fall, aggregate emissions targets decline, or output prices rise, firms have an incentive to bank, rather than borrow, permits. One prominent example is the Sulfur Allowance Program, where the early (banked) emissions reductions clearly lowered concentrations.

• When market failure in allocating emissions reductions over time is likely, it can be corrected by elements of program design. Specific examples of design modifications include: eliminating borrowing, introducing increasingly more stringent emissions targets, or using an intertemporal trading ratio that is not equal to one for allowing banked or borrowed permits to offset emissions.

• Existing tradable permit schemes differ considerably in how they treat banking or the role of forward markets. No existing system is fully temporally fungible. Older pollution control programs had a more limited approach. The Emissions Trading Program allowed banking but not borrowing. The Lead Phase-out Program originally allowed neither but part way through it allowed banking. The Sulfur Allowance Program has banking but not borrowing, and RECLAIM has an overlapping timeframe for compliance that is equivalent to a highly restricted banking and borrowing system. Banking of excess reductions for future years is allowed within the first phase of the European Union's emissions trading system, but banking between the 2005–2007 start-up phase and the 2008–2012 commitment period is at the discretion of member states.

• Ex post evaluations of several programs have revealed that the banking provisions have been quite important in terms of both saving costs and promoting earlier reductions.

## *Controlling Short-Term Concentration Peaks*

• Meeting short-term ambient standards cost-effectively means controlling the timing as well as the quantity of emissions. Two different types of permit systems are needed to cost-effectively meet two different types of situations. Periodic permits can be used to control short-term pollution peaks caused by regular, anticipated seasonal or diurnal variations in meteorological conditions. Episode permits can be used to control pollution during those rare, but potentially devastating, thermal inversions that can be anticipated only a day or so in advance.

- Though few studies have incorporated the temporal aspects of pollution, the available evidence suggests that significant cost savings may be possible from both periodic and episode control permits. These studies also find that a constant control policy based on a worst-case condition frequently is not sufficient to avoid violating the ambient standards.

- While older emissions trading programs took no account of seasonal factors in controlling pollutants, the NBP represents a significant change. Under the NBP, EPA established an ozone-season $NO_x$ budget for each affected state. Recognizing the importance of seasonality to ozone formation, this budget caps emissions from May 1 to September 30. The Chicago Emission Reduction Market System now has adopted this approach as well.

- While establishing controls only in seasonal periods could increase emission loadings significantly in periods other than peak periods, this emissions shifting can be countered through the use of annual average standards. Additional procedures would have to be established to prevent emissions from tall stacks escaping this annual average emissions accounting system. With this protection in place, cost-effective temporal control could be pursued without incurring large increases in emission loadings.

- To be cost-effective, episode control policies need to identify in advance those sources that can cut back relatively cheaply on very short notice. This can be accomplished by implementing an episode permit system in those areas plagued by episodes. Although dated and limited, the available evidence suggests that such an approach could reduce costs considerably.

## Notes

1. See, for example, Larsen (1971).

2. The standards in the United States can be found at: http://www.epa.gov/air/criteria.html.

3. In the United States, this apparently has happened. Henderson (1996) shows that localities in many nonattainment areas have improved short-term average readings without reducing total emissions.

4. Cronshaw and Cruse (1996) point out that if one or more firms are subject to profit

regulation, decentralized decisionmaking will not necessarily minimize costs. In general, they show that when some firms are subject to profit regulation, the permit price will exceed the marginal abatement cost by a differential that represents the increase in operating profit that a regulated firm would be allowed to earn to cover the cost of permit purchases.

5. Note the similarity to the necessary condition for allowing exchange rate trading in a zonal emissions trading market that was discussed in the previous chapter.

6. Forward markets allow the purchase of future allowances before their use date. For example, permits dated 2010 could be purchased in 2005. This is different from borrowing in that permits acquired in a futures market cannot be used until their designated date.

7. RECLAIM facilities are regulated in two cycles that have a six-month overlap. The compliance period for sources in the first cycle is January 1 through December 31, while second-cycle firms have a compliance period that runs from July 1 through June 30.

8. Note the similarity here of these temporal zones with the previously discussed spatial zones. In both cases, permits in different zones are treated differently.

9. This result can be seen most easily in O'Neil (1980, *129* and *141*). This provides a more complete description of the same model employed in O'Neil (1983).

# 6

# The Initial Allocation

Emissions trading approaches to pollution control involve two phases: (1) an initial allocation of permits before any trading takes place; and (2) an organized market or series of trading rules that allow permits to be transferred from one source to another. This chapter deals with the first of these important implementation steps.

In theory, the initial allocation should not matter much to cost-effectiveness. Any initial allocation would result in the same post-market allocation (after trading) and that allocation would be cost-effective.

In practice, it turns out the initial allocation matters a great deal, not only in terms of its impact on the fairness of the program but also on its cost-effectiveness. The initial allocation process also turns out in many emissions trading systems to be the most controversial aspect of the implementation process.

The flexibility offered by the many initial allocation possibilities is a double-edged sword. On the one hand, it allows the control authority wide latitude in its ability to pursue a just or fair distribution of the costs and benefits. On the other hand, this flexibility can trigger a political struggle in and of itself, with a majority coalition channeling the lion's share of the benefits to itself and a disproportionate share of the costs to a reluctant minority.

This chapter defines and explores the possibilities for the initial allocation of permits and uses this framework to understand choices made by current programs as well as to examine alternatives to those choices.

## Initial Allocation Approaches

After a careful survey of several specific applications of tradable permit systems in several different settings, Raymond (2003) demonstrated that the initial allocation is one of the most important and controversial parts of the process. The detailed description of the process for deciding the initial allocation of allowances in the Sulfur Allowance Program by Ellerman et al. (2000) supports this view. Not only do participants typically care deeply about the outcomes, but that passion results in a considerable amount of resources, both private and public, being devoted to designing and influencing the design of the initial allocation process.

In addition to affecting the likelihood that a program will get implemented, perceptions of fairness may affect program effectiveness as well. Some evidence, for example, suggests that enforcement may be both easier and more effective in programs that are perceived by participants to be fair (Hatcher et al. 2000).

## The Menu of Possibilities

The four possible methods for allocating initial entitlements are:

- random access (lotteries);
- first come, first served;
- administrative rules based upon eligibility criteria; and
- auctions.

All four of these methods have been used in one tradable permit context or another. Both lotteries and auctions frequently are used in allocating hunting permits for big game. Lotteries are more common for allocating permits among residents, while auctions are more common for allocating permits to non-residents. First come, first served was common historically for water and for harvesting fish, especially in conditions of abundance.

This chapter will focus on the two methods of most importance to emissions trading: administrative rules and auctions. The former is the dominant choice of existing programs, while the latter, though rare in practice, generally is considered by economists to have the most desirable efficiency properties.

## Administrative Rules

The most common basis for an administrative initial allocation of permits in all tradable permit systems, not just emissions trading, is to base the initial allocation on some combination of past use and equity norms. This approach typically is referred to as "grandfathering" because it not only takes past use into consideration but typically applies only to existing sources. New sources generally must either purchase permits from existing sources or be allocated permits on some other basis, such as by rule or negotiation.

Under administrative rules, the allocation commonly is based upon emissions or output. Mirroring the command-and-control standards with the initial allocation so as to make the adjustment costs small makes the transition easier but not easy. Most command-and control standards are based upon allowed emission rates, such as emissions per unit of fuel consumed, not emissions per se. By contrast, permit allocations are defined in terms of a specific quantity—typically emissions.[1] The shift from a technology and rate-based program to a mass/trading-based program changes the nature of the allocation. In particular, converting allowed emission rates to emissions requires specifying an authorized activity level, such as the allowed amount of fuel use, something typical emissions standards do not do. How this conversion is handled can have quite an effect on the initial allocation.

How initial allocations are designed in practice can be illustrated by how they played out in specific programs.

Under the Kyoto Protocol, the baseline for measuring emissions reductions was established by rule (1990 emission levels), but the amount of emissions reduction from that baseline assigned to each country was decided by negotiation. This negotiation resulted in very different allocations among the countries, ranging from Iceland's 10% increase to the European Community's 8% decrease.

In the Sulfur Allowance Program, the initial allocations were based roughly on two factors: fuel use and emissions rate per unit of fuel use. The fuel use factor was based on historical use but the emissions rate was not. It was based on an agreed-upon emissions rate that was applied uniformly to all covered emitters. As Ellerman et al. (2000, *41*) demonstrated, however, deviations from a strict application of these factors were common; negotiation was an important element of the process.

The Sulfur Allowance Program also included a ratchet clause. A ratchet clause ensures that any negotiated initial distribution is consistent with a previously stipulated cap. Frequently, negotiations over initial allocations result in a more generous outcome for emitters than originally envisioned, resulting in a situation where the sum of the negotiated allocations exceeds the cap. With a ratchet clause in place, this sum is reduced to the cap, usually by proportional reduction. In other words, if the sum of negotiated initial allocations turns out to be 120% of the cap, every actual allocation is then defined as the negotiated allocation divided by 1.20.

In the RECLAIM market, starting allocations were calculated by multiplying the maximum throughput during the period of 1989–1992 by equipment-specific emissions factors that related emissions to throughput. The equipment-specific emissions factors reflected the emissions reductions required by adopted district rules through December 31, 1993. The intent of selecting the highest throughput year was to replicate what would have been the facility's emissions in 1994 had it not been for the recession. Since previous command-and-control rules in the RECLAIM area did not place a cap on mass emissions (they controlled emission rates), allowed emissions could have reached the stipulated level, depending on the economy (U.S. EPA 2002).

Since, in contrast to the Sulfur Allowance Program, these initial allocations were not ratcheted down, in the early years of the RECLAIM program (1993–1999) most companies were allowed emissions levels that exceeded the levels that would have occurred under command-and-control. According to the South Coast Air Quality Management District's annual RECLAIM audit in 2000, the excess emissions authorized by this formula were 14,813 tons in 1994 and 10,267 tons in 1995.

One advantage of output-based gratis allocations is that the allocations can, in principle, be updated as output levels change. Thus, an electric utility that expands its output level relative to others automatically would receive a larger share of subsequent permits. Although they are more flexible, updated output-based allocations generally are neither efficient nor cost-effective since they provide an implicit subsidy to output. In contexts where tax distortions are important, however, the evidence suggests that they can hold their own with historic emissions-based allocations (Fischer and Fox 2004).

## Auctions

An auction provides an alternative way to handle the initial allocation. In an auction, buyers are pitted against one another to ensure that the auctioned item goes to the user that values it most. Although many different types of auctions exist (Vickrey 1961), for our purposes one crucial distinction is between auctions that produce revenue for the government and those that do not.

Since these permits are a creation of the government, it is normal to think of the government as deriving all the revenue from their sale. From a conceptual point of view, an auction where the revenue flows to the government is similar to a tax on emissions. These payments represent an additional financial burden on emitters over and above the cost of paying for the emissions abatement.

Not all auctions of permits, however, produce revenue for the government. Zero-revenue auctions, originally proposed in an emissions trading context by Hahn and Noll (1982, *141*), are a means of deriving the benefits of an auction without extracting large payments from the sources. Essentially all collected revenues are rebated to the sources, so the net transfer to the government is zero.

One version of a zero-revenue auction has been adopted in the Sulfur Allowance Program. Each year, the EPA withholds a bit under 3% of the allocated allowances and auctions them in an exchange run by the Chicago Board of Trade. These withheld permits are allocated to the highest bidders, with successful buyers paying their bid price. The proceeds are refunded on a proportional basis to the utilities from which the allowances were withheld.

Private allowance holders also may offer allowances for sale at these auctions. In the private sales, potential sellers specify minimum acceptable prices. Once the withheld allowances have been disbursed, EPA then matches the highest remaining bids with the lowest minimum acceptable prices on the private offerings and matches buyers and sellers until all remaining bids are less than the remaining minimum acceptable prices.

Notice that this auction does not replace the initial allocation procedure described above. Rather, it takes place after that distribution and serves mainly to ensure a steady flow of permits for sale (as a check against the possibility of hoarding) and to provide a transparent source for permit prices (to lower transactions costs).

One characteristic that differentiates auctions and administrative rules based upon historic experience is the different implicit ethical logic that lies behind the two baseline specifications. For auctions, the implied baseline control responsibility is for all sources to eliminate all emissions; when they fall short of that mark they must pay for the difference. Sufficient permits must be purchased from the government to cover the difference between complete and actual control; these deviations impose a financial liability on the emitter—the payments to acquire permits for any emissions. For existing sources, grandfathering implicitly assumes that emitters are entitled to emit specified levels as determined by the initial allocation; only emissions over those levels incur a financial liability in the form of payments for additional permits.

Another differentiating characteristic involves the tendency for administrative rules that focus on each individual source to over-allocate quotas in the initial years to enhance the political feasibility of the system. As we have seen with RECLAIM, the absence of a ratchet mechanism can result in a less stringent cap, at least initially. In the climate change case, a primary concern has been "hot air" (den Elzen and de Moor 2002), which is the part of an Annex I country's assigned amount of emissions that is likely to be surplus that was "earned" without any additional efforts to reduce emissions. Because assigned amounts are defined in terms of 1990 emissions levels, for some countries (most notably Russia and Ukraine) economic contraction has resulted in substantially lower emissions levels. As Maeda (2003) demonstrates, this choice of initial allocations could confer market power on those countries.

# Comparing the Allocation Approaches

## *The Impact on Implementation Feasibility*

Under virtually all implemented tradable permit programs, including, but not limited to, emissions trading, sources get free allocations of rights rather than having to pay for them. They only have to purchase additional permits that they may need over and above the initial allocation, as opposed to purchasing all permits in an auction market. Except in the EU ETS, new sources typically have to pay for permits regardless of the allocation method—either by direct purchase from sellers or though an auction.

How large would be the financial burden imposed by a requirement that all sources pay for all permits? Two measures of financial burden are relevant: the total regional financial burden imposed on all sources and the distribution of this burden among individual sources. The former sheds light on the significance of permit payments to control expenditures among the various approaches, while the latter examines the effects of the various distribution rules on individual sources, information that is helpful in discerning the additional costs those sources would face.

### Regional Financial Burden

The regional financial burden is defined as the sum of control costs and permit expenditures for all sources in the region. To show how the various methods of initially allocating credits affect regional financial burden, it is convenient to express this definition symbolically as

$$\text{regional financial burden} = \beta L + \alpha E$$
$$\beta \geq 1.0; -1 \leq \alpha \geq +1$$

where $L$ is the minimum possible control cost for meeting the ambient standard, $E$ is the maximum expenditure possible on permits, $\beta$ is the ratio of the control cost for the approach being considered to the minimum control cost for the pollution standard, and $\alpha$ is a coefficient that expresses the degree of financial participation of the government.

Since appropriately defined emissions trading systems are cost-effective ($\beta = 1$), differences in regional financial burden are due to the value of $\alpha$. For auction markets, $\alpha = 1$ because sources bear the entire permit expenditure burden collectively; every source has to purchase sufficient permits to legitimize its uncontrolled emissions. The grandfathering and zero-revenue auction options imply $\alpha = 0$ since for all sources taken together permit payments equal permit receipts.

Although from Chapters 2 and 3 we know that an appropriately defined emissions trading system would yield lower control costs than the command-and-control approach, total expenditures (including permit expenditures) may or may not be higher. $\beta$ is larger for command-and-control, but as long as $\alpha$ is positive, the payments for permits may more than offset the control cost savings. Furthermore, for an inappropriately designed permit system, such as when an emissions permit system is used to control non-uniformly mixed assimilative pollutants, it is not the case that control costs are minimized. Inappropriately designed emissions trading approaches can lead to larger finan-

**TABLE 6-1.** The Size of the Potential Regional Burden

| Study | | Ratio of total financial burden to CAC abatement cost | |
|---|---|---|---|
| | | Ambient permit | Emission permit |
| *Non-uniformly mixed assimilative pollutants:* | | | |
| Roach et al. (1981) | Sulfur dioxide | n.a. | 0.59 |
| Atkinson-Tietenberg (1984) | Particulates | 0.42[a] | 0.67[b] |
| Hahn and Noll (1982) | Sulfates | n.a. | 1.09 |
| Seskin et al. (1983) | Nitrogen oxides | 0.10 | 6.08 |
| Krupnick (1983) | Nitrogen dioxide | n.a. | 4.36 |
| McGartland (1984) | Total suspended particulates | 0.66 | n.a. |
| Spofford et al. (1984) | Sulfur dioxide | n.a. | 2.39 |
| | Particulates | n.a. | 0.35 |
| *Uniformly mixed accumulative pollutants:* | | | |
| Palmer et al. (1980) | Chlorofluorocarbons | N.A. | 8.4 |

*Definitions*: CAC = command-and-control; n.a. = Not available; N.A. = Not applicable.
*Sources*: Fred Roach, Charles Kolstad, Allen V. Kneese, Richard Tobin, and Michael Williams, "Alternative Air Quality Options in the Four Corners Region," *Southwestern Review* 1, no. 2 (1981): 44–45 (table 3); Scott E. Atkinson and T. H. Tietenberg, "Approaches for Reaching Ambient Standards in Non-Attainment Areas: Financial Burden and Efficiency Considerations," *Land Economics* 60, no. 2 (1984): 155 & 157 (tables 1 and 2); Robert W. Hahn and Roger G. Noll, "Designing an Efficient Permits Market," in Glen R. Cass et al., eds., *Implementing Tradeable Permits for Sulfur Oxide Emissions: A Case Study in the South Coast Air Basin* (vol. 1), a report prepared for the California Air Resources Board by the Environmental Quality Laboratory of the California Institute of Technology (1982), 106, 110; Eugene P. Seskin, Robert J. Anderson, Jr., and Robert O. Reid, "An Empirical Analysis of Economic Strategies for Controlling Air Pollution," *Journal of Environmental Economics and Management* 10, no. 2 (1983): 120; Alan J. Krupnick, "Costs of Alternative Policies for the Control of $NO_2$ in the Baltimore Region," unpublished Resources for the Future working paper (1983), 22 (table 4); Albert Mark McGartland, "Marketable Permit Systems for Air Pollution Control: An Empirical Study," Ph.D. dissertation, University of Maryland (1984), 74a (table 5.1); Walter O. Spofford, Jr., Clifford S. Russell, and Charles M. Paulsen, *Economic Properties of Alternative Source Control Policies: An Application to the Lower Delaware Valley*, unpublished Resources for the Future discussion paper D-118 (1984), 4.102 & 5.100 (tables 4.20 and 5.20); Adele R. Palmer, William E. Mooz, Timothy H. Quinn, and Kathleen A. Wolf, "Economic Implications of Regulating Chlorofluorocarbon Emissions from Nonaerosol Applications," report #R-2524-EPA prepared for the U.S. Environmental Protection Agency by the Rand Corporation (1980), 131.

cial burdens than the command-and-control approach because they involve financial outlays on permits (that command-and-control does not), and because $\beta > 1$ for both approaches.

The last two columns of Table 6-1 provide a direct comparison of the financial burdens of command-and-control and emissions trading for a variety of pollutants and regions when the credit expenditures are those that would result from an auction market ($\alpha = 1$).

Consider first the ambient permit column. Though only three estimates are available, all indicate that total compliance costs are lower with this permit system than with a command-and-control approach. If these studies turn out to be representative for this type of permit market, the estimated savings in control costs commonly exceed the estimated permit expenditures. Although this is an intriguing result, its policy significance is diminished by the complexities associated with an auction in ambient permits and the small number of studies involved.

In contrast to the results for ambient permits, financial burdens for emissions permit auctions frequently are higher than the command-and-control financial burden. Five out of the eight air pollution studies represented show financial burdens to be higher for emissions permits. In part, this is due to the higher control costs associated with the inherent over-control involved when using this approach to reach the ambient standards.

In those cases where the abatement costs are higher for the command-and-control approach than for emissions trading, a ratio of 1 or greater guarantees that the revenue from the auction would be sufficient to compensate sources so that they would be no worse off financially than under command-and-control. If the ratio is greater than 1, after compensating sources they will have revenue left over for other purposes. In cases where the emissions trading system results in higher abatement costs (for example, when an emissions permit system is used when an ambient permit system would have been appropriate), the revenue will not necessarily be sufficient.

Due to the small number of studies involved these results are hardly definitive. Nonetheless, they do provide a benchmark for subsequent studies to support or refute.

## The Distribution of Source Financial Burden

The preceding examination of regional financial burden sheds considerable light on whether the average source is better off under the auction system or the command-and-control approach. However, it also masks a considerable amount of variability among sources. Some sources will bear a lower financial burden than average and others will bear more. In other words, a ratio of 1 may guarantee that, collectively speaking, the sources are as well off financially under an auction version of emissions trading as under a command-and-control approach, but it does not mean each source will be as well off.

How serious is this unequal distribution of the burden? In his study of the control of aircraft noise, Harrison (1983, *114*) concludes that the extra payments to the government involved in an auction market would create greater cost disparities among airlines than the traditional approach.[2] He also notes, however, that the disparities would not be particularly great among close competitors. Cargo airlines, which typically operate the noisiest aircraft at night when the sensitivity to noise is the highest, would purchase the most permits. However, cargo airlines operating similar planes on similar routes would face similar cost increases; little competitive edge would be gained or lost by direct competitors as a result of these payments. He speculates, but does not perform the calculations to test his speculation, that variability among passenger airlines might be a more serious problem.

Palmer et al. (1980) also examine the interindustry and interfirm effects from using an auction market to control chlorofluorocarbons (Figure 6-1). Though this particular example is somewhat unusual because of the unusually large permit expenditures that would be involved, the variability of costs introduced by an auction market is striking.

One aspect of this figure that is particularly noteworthy is the "other" category. This category contains some product areas (rigid insulating foams, liquid fast freezing, and sterilants) where the authors find that no control methods would be introduced, even in an auction market. For these product areas, the only expense is incurred in purchasing permits and it represents a very large outlay. The manufacturers of these product lines could be expected to be unusually vociferous in their opposition to an auction approach and unusually warm in their support of a traditional regulatory approach that, in all likelihood, would place no control requirements on them at all.

The tendency for auction markets to increase cost disparities is not perfectly general. Lyon (1982) shows that firms with low marginal control costs will prefer uniform command-and-control regulations, while firms with high marginal control costs may or may not prefer auction markets. If in an auction market the control cost savings to the high-cost source are large enough, its cost disadvantage to the low marginal cost source would be reduced relative to the command-and-control approach in any move from command-and-control to an auction approach.

The discussion so far has focused on the distribution of financial burden among sources when the location of the sources is not a factor. The results of the Harrison study, for example, were in terms of a uniform marginal cost of control for each effective perceived noise decibel,[3] whereas the Palmer et al. study was based on a uniform marginal cost of controlling chlorofluorocarbon use in the United States.

When location is considered, either by introducing an auction market for ambient permits or by establishing different emissions permit auction markets in different local areas, another source of cost variability is introduced. Sources in geographic areas requiring higher marginal control costs would face higher financial burdens than direct competitors in areas with lower marginal control costs simply by virtue of their location.

Existing sources are particularly vulnerable because their location was determined before the system was implemented. Unlike new sources, they cannot pick a location to minimize costs. Although this cost differential would have in

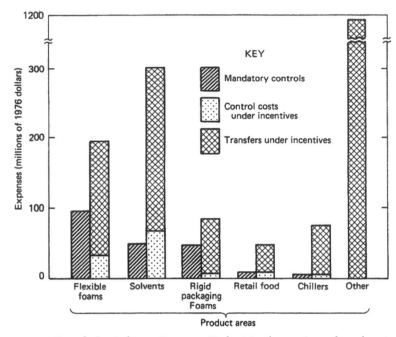

**FIGURE 6-1.** Cumulative Industry Expenses Under Mandatory Controls and an Auction Market for Permits

all likelihood been unanticipated by the existing sources, it could have a major impact on their market shares. Short of moving, there is little existing sources can do to unburden themselves of this competitive disadvantage. This is the flip side of the location coin. While considering the location of the emission provides the proper incentives for new sources to locate in low-damage areas, it also imposes higher costs on those sources that have already located in what turns out to be a high marginal control cost area, diminishing their competitiveness in the process.

One way to alleviate these side effects of auctions is to eliminate the transfer to the government, either by grandfathering or using a zero-revenue auction. For a given distribution rule and set of permit prices, these two approaches would yield the same distribution of financial burden among sources.

Other aspects of the initial allocation procedure also can raise competitiveness concerns. Negotiations, however, provide some "wiggle room" to lessen the problem. In the EU trading system, the cap for the sources included in the ETS was decided by each country subject to review by Brussels that the cap was consistent with achieving the EU cap on all emissions and that the ETS sector cap be binding. Although this almost guarantees that similar sources in different countries would be treated differently, given the difficulties of forecasting emissions and the many ways in which the national cap could be shared between the ETS and non-ETS sectors, each member state in fact had considerable leeway in setting the national cap.

## The Distribution of the Household Financial Burden

How does the burden of pollution control, as mediated by emissions trading, affect households? The answer to that question depends crucially on the form of the analysis.

One early study (Harrison 1994) uses a partial equilibrium analysis to estimate the burden imposed by the RECLAIM program on the people of the Los Angles area. His analysis suggests that because RECLAIM was estimated to reduce control costs by some 40% compared to a command-and-control approach with similar results, it would result in lower product prices. Given that the poor spend more (and save less) than the wealthy, these smaller price increases would benefit the poor more than the rich. Harrison also finds that RECLAIM was estimated to enhance employment relative to a command-and-control alternative and higher employment benefited the poor more as well since they depend more on wage income than returns to capital.

Dinan and Rogers (2002) use general equilibrium modeling to study the distributional effects of a program to reduce U.S. carbon emissions. In their model, product prices are higher with emissions trading when the permits are freely distributed, not lower. In their framework, permit rents are capitalized in higher firm equity values because firms receive an asset with market value for free. Acting like a binding production quota, the permit cap reduces output below free market levels, resulting in an increase in product prices, and firm profits are increased. Ultimately in this framework, permit rents accrue to households in the form of dividends or capital gains. To the extent that wealthy households receive a greater share of their income from capital than poor households, the creation of emissions trading scarcity rents is regressive.

Dinan and Rogers' (2002) results show that distributional effects hinge crucially on whether permits are grandfathered or auctioned and whether revenues from permit auctions or from indirect taxation of permit rents are used to cut payroll taxes, corporate taxes, or provide lump-sum transfers. They find that the lower income households would be worse off when the permits are freely distributed but better off in both absolute terms and relative to high income households when permits are auctioned off with revenues returned in equal lump-sum rebates for all households.

Similar results were obtained by Parry (2004). He examines the incidence of emissions permits, among other control instruments, to control power plant emissions of $SO_2$, carbon, and $NO_x$. He finds that using freely distributed emissions permits to reduce carbon emissions by 10% and $NO_x$ emissions by 30% can be highly regressive.

These differing results between partial and general equilibrium analysis deserve two explanatory comments. First, they clearly depend on what happens to product prices when an emissions trading regime is imposed. If freely distributed permits do indeed result in higher product prices, as theory leads us to believe, then the general equilibrium approach is more likely to be right. If they result in lower product prices, the partial equilibrium approach would be closer to the truth. To the author's knowledge, no current empirical study sheds light on this question.

Second, the superior distributional results for auctioned permits depend crucially on how the revenue from the auction is distributed. Since only the EU ETS authorizes initial allocation auctions, and then only for possible future use, this approach is so new that forming a reasonable judgment about how a rebate system would work in practice is difficult.

In the absence of evidence on these two points, the distributional consequences of emissions trading remains an open question.

## Implementation Feasibility

Financial burden considerations affect the attractiveness of alternative policy approaches to environmental protection from the point of view of the various stakeholders. To the extent that stakeholders can influence policy choice, using initial allocations based upon administrative rules may increase the feasibility of implementing emissions trading systems (Svendsen 1999; Raymond 2003).

Interestingly, however, for at least some applications of emissions trading where the cost savings are substantial, empirical evidence suggests that the amount of revenue needed to hold users harmless during the change is only a fraction of the total revenue available from auctioning (Bovenberg and Goulder 2000). The implication is that while a small portion of the revenue from auctioned permits would be rebated to sources to ensure political feasibility, the rest could be used to reduce distortionary taxes, thereby lowering program cost. Allocating all permits free of charge is therefore not inevitable in principle, even if financial burden proves to be an important consideration.

Political feasibility, however, may not be the only stumbling block to the use of auctions. The ethical importance of the initial allocation may be important in other ways. Raymond's (2003) detailed review of the initial allocation processes for three major tradable permit programs concludes not only that equity norms played a large role in crafting the initial allocation in these cases but also that applying these norms is much more complicated than simply relying upon prior use. His analysis further suggests that in terms of prevailing equity norms, auctions may have a tough time gaining a foothold in initial allocations despite their attractiveness from an efficiency point of view.

# Cost-Effectiveness Implications of the Initial Allocation

The conventional wisdom held that the initial allocation would have no impact on cost-effectiveness. Any cost-ineffectiveness associated with the initial allocation would be eliminated by subsequent trading. In practice, this turns out to be a considerable overstatement. In several circumstances, the degree of cost-effectiveness achieved by emissions trading is dependent upon the initial allocation not independent of it.

## The Double Dividend Issue

Recent general equilibrium analysis examines how the presence of preexisting distortions in the tax system affects the efficiency of the chosen pollution control policy instrument in general and the role of the initial allocation in particular. This makes it clear that not only can emissions permit systems drive up the output price from polluting industries—which can exacerbate other economic distortions from the tax system, thereby increasing the cost of emissions trading policies—but the ability to recycle the revenue from the sale of permits (rather than give it to users) can enhance the efficiency of the system by a large amount. Central-case estimates suggest that the net effect of such interactions is to increase the cost of emissions trading slightly when all of the permits are auctioned but by much more if they are not (e.g., see Burtraw et al. 1998; Parry et al. 1999; Goulder et al. 1999). That work, of course, supports the use of auctioned permits rather than free distribution.

## Differential Treatment of New Sources

A second concern about cost-effectiveness arises from the differential treatment of existing and new sources. With the exception of national plans in the EU ETS, new firms typically have to purchase all permits, while existing firms get an initial allocation free. Thus, the free distribution system imposes a higher financial burden (the sum of abatement costs and permit costs) on new sources than on otherwise identical existing sources. Empirical work on this effect suggests that this differential treatment has retarded the introduction of new facilities and new technologies by reducing the cost advantage that would otherwise accrue from building new facilities embodying the latest innovations[4] (Maloney and Brady 1988; Nelson et al. 1993).

## Market Power

As demonstrated in the next chapter, a free-distribution initial allocation also can introduce market power by granting firms an excess of permits that they can use to manipulate the market.

## Strategic Considerations

Because permits are valuable and a free-distribution initial allocation provides an opportunity to capture this value, basing the initial allocation on prior use can promote inefficient strategic behavior. When the initial allocation is based upon historic use and users are aware of this aspect in advance, an incentive to inflate historic use to qualify for a larger initial allocation is created (Berland et al. 2001). This strategic behavior could intensify air quality deterioration in the period before emissions trading is initiated.

## Transactions Costs

The conventional premise that the initial allocation does not matter in achieving cost-effectiveness assumes the absence of transactions costs. In the presence of transactions costs, the post-trade allocation will not only depend on the initial allocation but will not, in general, be cost-effective.

## Regulated Markets

While theoretical models frequently assume unfettered markets, in fact many permit markets involved participants that are heavily constrained by regulations. When the effect of these regulations is to diminish the incentive to minimize costs or the incentive to trade, the initial allocation may prove to be more durable than normally would be the case. These observations seem particularly relevant for markets that involve regulated utilities (Coggins and Smith 1993; Fullerton et al. 1997; Rose 1997).

Some empirical confirmation is beginning to emerge of the general point that in practice initial allocations affect cost-effectiveness. Using panel data from Southern California's RECLAIM program, Fowlie and Perloff (2004) find not only that initial allocations are a statistically significant determinant of firm-level actual emissions but that the relationship is stronger among firms with relatively high transactions costs. Specifically, they use two approaches to identify firms that are likely to have relatively high transactions costs: small firms and firms that had not previously participated in the market. Although they find no clear evidence of a transactions cost effect with respect to firm size, they did find one for firms that had not previously participated in the market.

For all of these reasons, the initial allocation matters and evidence suggests that it can matter a great deal. It matters not only because both financial burdens and equity norms play an important role in the process of creating the allocations but also because in some circumstances the degree of cost-effectiveness actually achieved by emissions trading is affected by the initial allocation.

# Summary

- In traditional theory, emissions trading can achieve a cost-effective allocation regardless of the initial allocation. That flexibility would at first glance appear to offer the opportunity to achieve both cost-effectiveness and fairness by using the initial allocation to resolve fairness issues. The absence of a tradeoff between these objectives is quite unusual in public policy circles.

- Recent general equilibrium analysis shows, however, that the presence of preexisting distortions in the tax system can affect the efficiency of the chosen pollution control policy instrument in general and the role of the initial allocation in particular. This analysis demonstrates that not only can emissions permit systems drive up the output price from polluting industries—which can exacerbate other economic distortions from the tax system, thereby increasing the cost of emissions trading policies—but the ability to recycle the revenue from the sale of these permits (rather than give it to users) can enhance the efficiency of the system by a large amount. The net effect of such interactions is to increase the cost of emissions trading slightly when all of the permits are auctioned but by much more if they are not. That work, of course, supports the use of auctioned permits rather than free distribution, a strategy that is rarely used in current emissions trading programs.

- In principle, the control authority has many options as to how it distributes the financial burden in emissions trading. Four possible methods for allocating initial entitlements are: (1) random access (lotteries); (2) first come, first served; (3) administrative rules based upon eligibility criteria; and (4) auctions. In practice, virtually all existing emissions trading programs are based at least in part upon administrative rules. Most of these use rules that are based partially on past use, particularly for determining activity levels, and equity norms, particularly for determining the emissions rate per unit of activity.

- One advantage of output-based gratis allocations is that the allocations can in principle be updated as output levels change. Updated output-based allocations generally are neither efficient nor cost-effective, since they provide an implicit subsidy to output, but in contexts where tax distortions are important they can hold their own with historic emissions-based allocations.

- Under emissions trading, any source required to control its emissions faces two types of financial burden: (1) control costs; and (2) permit expenditures. The empirical evidence suggests that permit expenditures are sufficiently large that sources typically would have a lower financial burden under the command-and-control approach than under an auction permit approach in the absence of any rebate of the revenue. This significant source of political opposition to auction markets apparently has undermined their acceptance in the past, though in at least some cases, rebates of only a part of the revenue would make the auction financial burden equal to that under a command-and-control approach. With carefully structured rebates, the financial burden of auctions does not need to be an impediment in the future.

- Studies of the control of carbon emissions show that distributional effects hinge crucially on whether permits are grandfathered or auctioned and whether revenues from permit auctions or from indirect taxation of permit rents are used to cut payroll taxes, corporate taxes, or provide lump-sum transfers. Generally, these general equilibrium studies find that lower income households would be worse off when permits are freely distributed but better off in both absolute terms and relative to high income households if permits are auctioned off with revenues returned in equal, lump-sum rebates for all households.

- Grandfathering and zero-revenue auctions provide two ways to allocate control responsibility without financial transfers to or from the government. Both require the control authority to specify some rule for distributing the baseline control responsibility.

- Case studies, especially of the RECLAIM program and the Kyoto trading system, reveal some tendency to over-allocate quota in the initial program years to gain political feasibility for the system. In the absence of a ratchet mechanism, this may result in a less stringent cap, at least initially, and can mean that a command-and-control approach could achieve more emissions reduction in the early years.

- In several practical circumstances, the degree of cost-effectiveness achieved by emissions trading is dependent upon the initial allocation, not independent of it. Specific examples of this dependence are illustrated by: (1) the double dividend issue, where recycling the revenue from auctions can reduce the welfare costs; (2) the differential treatment of new sources under grandfathering; (3) creating market power via the initial allocation; (4) strategic considerations, where firms increase their emissions to qualify for more permits; and (5) transactions costs, where trading is sufficiently inhibited that the ability of the market to overcome any cost-effective deficiencies in the initial allocation is precluded.

# Notes

1. Not all programs are defined in terms of emissions. The Lead Phase-out Program, for example, was defined in terms of the lead content of refined gasoline.

2. Harrison bases this conclusion on a noise charge rather than an auction market. Given the mathematical equivalence of the two approaches, his conclusion holds for auction markets as well.

3. Aircraft noise is measured universally in terms of effective perceived noise decibels. This scale modifies the basic decibel scale by accounting for how people judge the noisiness of aircraft takeoffs and landings. High-frequency noises, for example, receive more weight.

4. The "new source bias" is, of course, not unique to tradable permit systems. It applies to any system of regulation that imposes more stringent requirements on new sources than existing ones.

# 7

# Market Power

Although emissions trading offers many advantages to authorities seeking to control pollution, it also has one potential disadvantage that is not shared by other pollution control policy instruments, such as emissions standards and emissions charges: permit markets potentially can be manipulated by those with market power.

This chapter explores what is known about this aspect of emissions trading. When can market power arise? What are its consequences? When does it become sufficiently burdensome that it offsets the comparative advantage of permit markets vis-à-vis other instruments?

Two different types of market power will be considered in the search for answers to these questions. The first type occurs when an aggressive source seeks to influence permit prices. What happens to the other sources in this type of market power is incidental, not central—power over the permit market is an end in itself.

In the second type, a source or coalition of sources attempts to leverage power in one market (either the output or the permit market) to gain an economic advantage in the other market. In this situation, a firm may act strategically in one market, gaining less than the maximum profit in that market, in order to produce more than offsetting profits in the other market.

It is necessary to differentiate these objectives for market power because each arises from a unique set of circumstances, has different consequences, and offers different mitigation possibilities.

## Permit Price Manipulation: Conceptual Models

Suppose that one or more firms (which we shall call price-setting firms) seek to exercise control over permit prices to reduce their financial burden. The extent to which they succeed depends on a number of factors, one of which is the initial allocation method.

## Traditional Auctions and Subsidy Auctions

**Traditional Auctions**. If a traditional auction market is used to distribute permits, all emitting sources would necessarily be seeking permits and the control authority would be the only seller. The ability of any source, or coalition of sources, to affect the prices paid for those permits would depend on the magnitude of its demand compared with the demands of the other sources. In a market with only one source, the source could acquire the permits at a negligible cost. Since any positive bid would be decisive, permit expenditures would be close to zero for this source, even in an auction.

The situation is more complex with multiple sources. Suppose that one particular firm wished to exercise control over prices, while all others were content to act as price takers. The price-setting firm, by articulating an artificially low demand for permits, could lower the price.

For simplicity, assume that only two sources are bidding for these permits. The first source is presumed to use its purchasing behavior to control the price, while the second source is presumed to be a price taker. By definition, the financial burden of the first (price-setting) source is the sum of its permit expenditures and its control costs.

The price-setting source would minimize its financial burden by equating the marginal expenditure on permits, taking the effect of further purchases on prices into account, to the marginal cost of control. Every additional permit purchased by the first source would drive the price higher, not merely for the additional permits, but for all permits. Therefore, to hold price down, the price-setting source must purchase fewer than normal permits, implying that its control costs would be higher than normal.

Several key insights flow from this simple conceptual model:

- Permit prices are lower in the non-competitive auction than in the competitive auction.

- The price-setting firm controls more emissions than it would if it were merely acting as a price-taker in a purely competitive market. Because total emissions from all sources would be the same in competitive and non-competitive markets due to the cap, the price-taking source would have to control fewer emissions in the non-competitive market than in the competitive market.

- Note that in this case the price-taking firm is a beneficiary of the price-setting firm's actions (needing to control fewer emissions), not a victim. Furthermore, in terms of reduced financial burden, the price-taking source benefits more than the price-setting source. The price-setting source has to control more emissions to lower the price, but the price-taking source benefits from the lower price without having to control more emissions.

- The non-competitive auction market allocation of control responsibility is not cost-effective; aggregate control costs to achieve the targeted level of emissions reduction are higher in non-competitive markets than in competitive markets.

To those envisioning the price-setting source as inflicting significant harm on other less aggressive sources, these findings may appear surprising. In this form of market power, control costs do rise when one source becomes a price-setter, but harm is not inflicted on other sources. However, this characteristic of market power does not carry over to other forms.

**Subsidy Auctions.**  In many ways, a subsidy auction (where the government asks firms to bid on emissions reductions and the government selects and funds the lowest cost reductions that meet the target) would be the mirror image of the traditional auction approach. It should, therefore, not be surprising that the presence of a price-setting firm in a subsidy setting produces results that are similar.

In a subsidy auction, the price-setting firm would try to raise the permit price above its competitive level by restricting the amount of emissions reductions offered to the government. Comparing a non-competitive subsidy auction allocation to a competitive subsidy auction allocation reveals that the non-competitive permit price would be higher, the price-setting source's control cost would be higher (since it offers fewer reductions for sale), the government's expenditures on securing the desired reductions would be higher, and the price-taking source would, once again, gain more than the price-setting source.

The conceptual foundations for this analysis also are helpful in isolating the factors that determine the degree to which a source can manipulate either type of auction for its own gain. One such factor is the relative importance of the price-setting source's demand for permits compared with the demand of the competitive, price-taking fringe. When the price-setting source's demand for permits comprises only a small proportion of total demand, manipulating its demand will have little or no influence on the price. Unless the source controls a sufficient proportion of the total demand, its efforts are for nil.

The shapes of the cost functions for both price-setting and price-taking sources also are important. The slope of the price-setting source's control cost function determines the degree to which it can gain from manipulating the price. Since the price-setting source secures all price changes by controlling more emissions, the cost of its strategy is determined by its marginal cost function. Higher marginal costs lower the gain from a given reduction in demand; the high cost of producing the compensating emissions reduction limits the ability to exercise market power.

The cost functions for the price-taking firms also are important. When the demand for permits by the price-taking firms is price inelastic, price-setting firms can have a larger effect on price without having to reduce their own emissions as much to compensate. Although the financial burden of all sources could be greatly affected by price manipulation in this case, control costs would not be affected.

Conversely, the power of any source to lower its financial burden would be diminished when the demand for permits by the price-taking firms is price sensitive. In this case, any given price reduction would trigger a larger increase in the demand for permits by these sources, causing a larger increase in the need for the price-setting source to control its own emissions, a costly outcome for the price-setting source.

Two other potential fears about price manipulation can be laid to rest:

- Regardless of the circumstances, as long as enforcement is effective, price manipulation should not affect air quality because, by design, the auctioned permits hold air quality constant. While price manipulation affects the distribution and cost of permit holdings, it does not affect total emissions.

- Furthermore, in either traditional or subsidy auctions, less aggressive price-taking firms are not harmed by price manipulation; they are benefited by lower financial burdens to an even greater extent than is the price-setting source. The only financial loser from this type of market manipulation is the government. It would receive less revenue from a manipulated traditional auction and would have to make larger payments in a manipulated subsidy auction.

### Free Distribution Initial Allocations

Traditional and subsidy auctions are only two forms of permit markets; however, most emissions trading programs do not involve either. Rather, as the previous chapter made clear, they involve initial allocations that distribute the permits free-of-charge on the basis of some kind of allocation rule. How does this form of initial allocation affect market power?

Hahn (1982) has shown that a free distribution initial allocation can influence the nature of the market power problem in significant ways. The most important of his results demonstrate that the potential for market power to be exercised is a function of the particular baseline allocation of control responsibility chosen by the control authority. Both the number of permits traded and the degree to which the aggressive source can dominate these trades depends on how the pre-trade control responsibility is allocated.

Several other conclusions follow from Hahn's (1984) theoretical analysis of the free distribution case:

- If the free distribution baseline allocation of permits turns out to be cost-effective, the existence of one or more price-setting sources would not raise total control costs. Because no trades would take place in this case, firms have no opportunity to exert power.

- Whenever a single price-setting source receives a baseline control responsibility either exceeding or falling below its cost-effective allocation, total control costs would exceed their minimum. When the baseline control responsibility falls below the cost-effective allocation, the price-setter can exercise power on the selling side and when it is above the cost-effective allocation, the price-setter can exercise power on the buying side of the market.

- As baseline control responsibility hypothetically is transferred from price-taking buyers to price-setting sellers, the permit price would rise and the number of credits retained by price-setting firms for their own use would increase.

- In a free-distribution market, the ability of any one source to affect permit prices is a function of its net demand for or net supply of permits (after considering its initial allocation), not the size of the source per se.

These results suggest that the flexibility that control authorities have in principle in allocating the financial burden could be substantially less in practice if market power is a potential problem. Deviations from the cost-effective control baseline could cause control costs to exceed their minimum level by opening price manipulating opportunities. The crucial question is how sensitive control costs are to these deviations. If they turn out to be insensitive, then the control authority's flexibility is not seriously jeopardized.

Hagem and Westkog (1998) extend Hahn's (1984) model by investigating how market power could influence permit trading over time. In their two-period model, they posit a situation in which the initial allocation allows one big firm always to be a seller of permits. All firms are assumed to have complete flexibility in banking and borrowing. Only the total amount of emissions between the two periods is controlled.

Their results suggest that market power does not interfere with the allocation over time. Specifically, both the competitive fringe and the monopoly seller choose reductions between the two time periods such that the present values of marginal abatement costs are equalized across time periods. The problem of market power in this case is manifested as a problem in the allocation of control responsibility across firms, not across time. Since the price-setter sells too few permits to artificially raise prices, its marginal abatement costs are smaller than those of the price-taking firms (as opposed to equalized as cost-effectiveness requires).

These results turn out to be sensitive to model specification. Unfortunately, the two-period model of Hagem and Westkog (1998) does not capture the full story of market power in a permits market with banking over a longer time horizon. Liski and Montero (2005) develop a model with a dominant firm and a fringe of small firms in a permit market that allows banking. Their model implies that market power is manifested as a problem in the allocation of responsibility across firms and across time. In fact, they find that in this setting, permit prices do not rise at the rate of interest throughout. As soon as fringe members exhaust their bank of allowances and the dominant firm is the only one left with a stock of allowances, which happens in equilibrium, allowance prices start rising at a rate strictly lower than the interest rate. This change in the differential behavior of the price increases in the two cases results in the misallocation of resources across time.

## *Market Power and Regulatory Compliance*

So far, we have assumed that firms all comply with the regulations. Suppose we now consider the possibility of noncompliance. Would the presence of market power ameliorate or exacerbate non-compliance behavior?

Van Egteren and Weber (1996) consider a case where firm emissions might exceed what is allowed by the permits. In this model, firms are audited with a certain probability and fined whenever cheating is discovered. Their chief finding is that when a firm has market power in the permit market, the initial allocation has a major effect not only on prices but also on levels of compliance for all participants in the permit market. In particular, they show that:

- When the dominant firm is compliant, increasing its initial allocation of permits, while imposing an equivalent reduction in the endowment of the price-taking, noncompliant fringe, leads to an unambiguous increase in aggregate violations and aggregate emissions.

- When the dominant firm is noncompliant, then the impact of allocating more via the initial allocation to the dominant firm has an ambiguous impact on aggregate violations and emissions.

The ambiguity results from the increase in the equilibrium price of permits caused by market power. By increasing the incentive to cheat, this price increase leads to an increase in violations and a decrease in emissions among the fringe. Although the dominant firm holds more permits in equilibrium, its violations decrease even though its emissions increase. Since violations and emissions for the two groups move in opposite directions, the overall impact on global compliance and emission levels depends on the specific circumstances.[1]

Malik (2002) extends the model of Van Egteren and Weber (1996) by investigating the efficiency consequences of this combination of non-compliance with market power. He shows that from an efficiency point of view, non-compliance by the small, price-taking firms is potentially desirable in the presence of market power; it reduces the negative impact of the restrictions imposed by the firm with market power by increasing their emissions and reducing their excessive compliance costs. Conversely, in the presence of non-compliance, some market power by the price-setting firm could be desirable, because it has an incentive to retire some permits and thus reduce some of the excess pollution that is emitted through noncompliance.

## Leveraging Power Between Output and Permit Markets

The previous section examined how the initial allocation of permits could give rise to market power, as well as how that market power could affect the outcomes of the permit market. In the models examined in that section, the permit market was assumed to be independent of other markets, most notably the output market.

Clearly output and permit markets need not be independent. Power in the permit market conceivably could be used to increase market share in the output market if some subset of firms were in both markets. How might this situation unfold?

Misiolek and Elder (1989) extend the Hahn (1984) model by assuming that one big firm acts as a price-taker in the output market but a price-setter in the permit market. Can this firm leverage its market power in the permit market so as to increase its market share in the product market?

Their results suggest that it can. Specifically, they find that the dominant firm seeking to leverage its power in the permit market always holds more permits than it would otherwise. Preventing rivals from holding these permits raises their production costs and thereby increases both the output price and the dominant firm's market share. In essence, considering output market effects provides an additional reason for firms with market power to hoard permits.

Innes et al. (1991) consider a slightly different case. Suppose a large firm has market power in both the output and permit markets but the price-takers in the permit market produce in a different output market. In this case, raising a rival's costs is not the means for maximizing profits in both markets but price manipulation is. Because the market power in the product market reduces the dominant firm's demand for permits, it provides another benefit from hoarding (the associated permit price manipulation).[2]

Sartzetakis (1997a) pursues a similar approach as Misiolek and Elder (1989) but assumes that two firms engage in imperfect (Cournot)[3] competition in the output market, while only one of them has market power in the permit market. His results show that in this case as well market power in the permit market can be leveraged to reduce competition in the output market.

Does the presence of market power overturn the traditionally strong case for allowing trading of emissions responsibilities? Compared to a command-and-control approach where no trading is allowed, does the introduction of trading always increase welfare even in the presence of market power?

Sartzetakis (1997b) examines this question in the context of a model that is similar to his model described above. His results indicate that allowing trading of emissions permits has two effects: (1) due to the competitive permit market, it minimizes the cost of emissions control effort by equalizing the marginal cost of abatement among firms; and (2) it redistributes production among firms both due to imperfections in product markets and to firm-specific differences in emissions control technologies. These effects have different welfare consequences.

Cost-minimization in the competitive permit market is clearly a welfare-increasing effect, while production redistribution could conceivably be welfare decreasing if the inefficient firms are the ones that gain market share. As the paper shows, however, the welfare-increasing effect dominates. Even in the presence of this form of market power, trading increases welfare.[4]

## *Market Power and Innovation*

Another aspect of market power involves its potential connection to innovation. Could innovation affect the likelihood or degree of market power? Would the existence of market power inhibit or intensify innovation incentives?

**The Effect of Innovation on Market Power.** Fischer et al. (2003) examine the first question in the context of a model where a single innovative firm is able to invent a new technology that lowers marginal abatement costs. Since the innovative firm needs fewer permits after the new technology is installed, the innovation potentially would provide some market power (due to the excess permits) in the permit market. In this case, the ability to obtain some market power provides an extra incentive to innovate.

**The Effect of Market Power on Innovation Incentives.** How does the presence of market power affect the relative advantages of various pollution control policy instruments in terms of their ability to promote innovation? Montero (2002a) studies the investment incentives created by tradable permits and two kinds of standards, emissions and performance standards. He models imperfect competition in both the output and the permit market.

Since in his model firms have market power in both markets, permit trading decisions are affected by a strategic effect in the output market. He finds that strategic considerations cause the firms to invest more in innovation under an emissions standard and a regime of auctioned permits than under a regime of free permits. An emissions tax, in contrast, has no strategic effect.

The explanation is that the strategic effect under emissions standards is always positive, in that a firm's innovation reduces its own costs but not those of its rivals, allowing the firm to increase output and profits. Under tradable (auctioned) permits, however, the strategic effect may be negative because a firm's investment in innovation spills over through the lower prices in the permit market (or permit auction), reducing its rivals' costs and thereby helping rivals to increase output.

## Ex Ante Simulations

While the directions of these effects are clear from theory, the magnitudes are not. How affected by the presence of market power are total control costs? Since the answer to this question is pivotal in assessing the significance of market imperfections, the evidence that flows from simulations is of some importance.

Unfortunately, few studies address this issue but three have. Hahn (1984) examined the sensitivity of control costs to changes in the initial allocation of control responsibility to one potential price-setting source, using data on sulfur dioxide control in Los Angeles. The results are reproduced as Figure 7-1.

The cost-effective initial allocation of allowable emissions is indicated as Q*. Control costs are flat in the neighborhood of that allocation. The one range of initial allocations in the figure that shows a large increase in control costs involves allocating a very large proportion of allowable emissions (or a small proportion of baseline control responsibility) to the price-setting source. In this case, the price-taking sources are all on very steep portions of their marginal cost functions, while the price-setting source is on a relatively flat portion of its marginal cost function. Add the fact that the price-setting firm, by virtue of its high initial allocation of allowable emissions, controls a substantial proportion of the credits available for sale, and it becomes clear why control costs are affected to such a high degree. The price-setting firm has the other sources over a barrel. Should they fail to acquire additional credits, their control costs would be so high as to threaten their continued existence.

Though it appears from Figure 7-1 that monopsony—created by allocating too few permits to the price-setting source—is not a problem, that conclusion may be sensitive to the specific model employed. In other work, Hahn and Noll (1982, *135–137*) examine in the context of sulfate control in Los Angeles the effect of choosing a control baseline in which one large source is required to control all emissions while other sources control less than their cost-effective amount. This construction creates a situation in which the large source is the only buyer, facing a number of credit sellers. Calculations of the increase in control cost due to this form of market power were performed for a number of cases involving different assumptions about natural gas availability and levels of desired air quality. These calculations showed the losses to be relatively small, ranging from zero to 10%, depending upon the case examined.

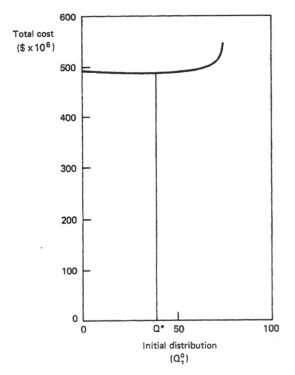

**FIGURE 7-1.** Total Annual Abatement Cost vs. Initial Permit Distribution

In another set of published data from the DuPont Corporation involving some 52 plants and 548 sources of hydrocarbons, Maloney and Yandle (1984) investigated the effects of cartelization of plants on the permit market. Assuming that all sources receive a proportional initial distribution of the permits based on their uncontrolled emissions, they calculate the effects on control costs if plants collude. Their analysis allows collusion to take place separately among buyers and sellers and allows the number of colluding plants to vary from 10–90% of the total number of plants buying or selling.

In general, these data support the notion that high degrees of cartelization are necessary before control costs are affected to any appreciable degree and that even high degrees of cartelization do not significantly erode the large savings to be achieved from permit markets. The 90% permit monopoly (achieved when the cartel controls 90% of all permits sold), for example, yields a 41% increase in control costs. Maloney and Yandle (1984) point out that the cost savings from even this severe market power situation, compared with command-and-control regulation, is still 66% (instead of 76%). In this study, the presence of market power does not seem to diminish the potential for cost savings very much. Even with market power, emissions trading seems to result in lower control costs than the command-and-control approach.

These data also support the notion that market power on the seller side is a more serious problem than market power on the buyer side, though the results do so indirectly. The number of plants selling permits (8) is estimated to be substantially fewer than the number of plants purchasing permits (44). This is a natural consequence of the proportional distribution rule used by Maloney and Yandle (1984), which favors sources with large economies of scale. Because the transactions costs associated with forming a cartel with a large number of small sources are significantly greater than those for forming one with a small number of large sources, proportional initial allocation rules make power on the seller side more likely than on the buyer side by creating a situation involving a few plants selling permits to a much larger number of buyers. Initial allocation rules are likely to create power on the buyer side only if they result in a few buyers facing a large number of sellers.

Given the paucity of data on air pollution cases, one early study that focuses on water pollution control is included in this review. de Lucia (1974, *124–131*) examined the effect of market power on permit prices and control costs. This study covered eight sources emitting two different water pollutants (biomass potential and biochemical oxygen demand) into the Mohawk River.[5] In these two simulations (one for each pollutant), the largest source is assumed to act as a price-setter, with the remaining sources acting as price-takers. In each simulation, the designated price-setting source was more than three times larger than the next largest source, accounting for more than 46% of the uncontrolled total biomass potential and more than 45% of the biochemical oxygen demand load placed on the river.[6] By arranging its bids, the price-setting source is assumed to manipulate the price so as to minimize its financial burden.

In the simulation of biochemical oxygen demand, the price-setting source only has a negligible effect on price and on the regional cost of controlling all sources. Its attempts to lower the price of permits were effectively thwarted by the cost of the additional control it would have to bear. Its control costs were so large as to prevent it from securing any gain from price manipulation.

In the simulation of biomass potential, the price-setting source does gain a somewhat lower financial burden than it would have if no source acted as a price-setter, but the effect on regional control costs is still negligible. They rise less than one-fifth of 1%.

de Lucia (1974) also measured the sensitivity of his conclusions by considering a third simulation involving a hypothetical market containing only two polluters: one price-taker and one price-setter emitting roughly the same uncontrolled levels of waste. In this simulation, permit price was found to be quite sensitive to market structure, falling in the non-competitive market to less than half its competitive level.

The most interesting finding from the point of view of assessing the seriousness of imperfections in auction markets concerns the effect of market power on control costs. Even when one source controlling roughly half the market aggressively manipulates auction market prices in its own interests, control costs rise by less than seven-tenths of 1%.

Though the available evidence is very limited, it is remarkably consistent in finding that control costs do not seem to be very sensitive to market manipulation. We have examined a very congenial setting for market manipulation, one involving few sources, low marginal control costs for the price-setting source, and high control costs for the competitive fringe. Despite finding circumstances where prices and total financial burden were affected dramatically, regional control costs were remarkably insensitive to market manipulation in all ex ante simulations.

# Results from Experimental Studies

An additional source of information about market power can be derived from the literature in experimental economics. In this sub-field of economics, researchers use human subjects and experimental methods to evaluate theoretical predictions of economic behavior. Most experiments of this type have been conducted in controlled laboratory settings, but recently interest has grown in field experiments.

## Efficiency and Price Manipulation

In one such experiment, Godby (2000) examines the effects of market power in both the permit market and the product market. In an experiment with a single seller, which could manipulate price, and ten fringe firms, which behave as price-takers, he finds that the experimental results are much closer to the prediction of the theoretical Hahn model than to the prediction of the competitive model. To be specific, his experiments find that the inefficiency caused by market power could exceed even the inefficiency caused by command-and-control. This, of course, is a very different result than obtained from the simulations discussed in the previous section.

## Controlling Power Through Market Design

Might some specific auction market designs be preferable when the possibility of market power looms? One of the earliest studies to examine this question looked at how well zero-revenue auctions would fare when the possibility of market power was present. Hahn (1983) designed a series of experiments involving three separate simulations using a total of 24 students. A different initial allocation of permits was used for each of the three simulations.[7] Each simulation was repeated 10 times to allow for learning to take place.

Hahn (1983) finds that even when the initial allocation deviated considerably from the least-cost allocation, the zero-revenue auction tended to converge to the least-cost allocation. Subsequent experiments by Hahn yielded similar results. This seems to confirm our earlier finding that auctions, even zero-revenue auctions, are less susceptible to market power than free-distribution allocations.

Does the type of auction matter? Muller et al. (2002) focused on another auction design—the double auction.[8] As the authors note, this particular design had been thought to be less susceptible to market power than other auction forms; the experiments were designed to test this conjecture.

In their experiments, they introduce market power on the seller or the buyer side by aggregating five sellers or five buyers, respectively. The main conclusion from their experiments is that the double auction design is not as robust to market power as conjecture would have us believe.

Cason et al. (2003) also report on a laboratory experiment to examine whether a dominant firm could exercise market power in a permit market organized using the double auction trading institution. The parameters in their experiment were designed to approximate the abatement costs of sources in a proposed tradable emissions market for the reduction of nitrogen in the Port Phillip Watershed in Victoria, Australia. The initial allocation of permits to sources were varied, so that in one treatment the seller of permits was a monopolist and in another treatment the market was duopolistic. They found that prices and seller profits were higher and efficiency was lower on average in the monopoly treatments compared with the duopoly treatments, but the differences were not substantial and were not statistically significant. Moreover, they found that prices, profits, and transaction volumes usually were much closer to the competitive equilibrium than the monopoly equilibrium.

Carlén (2003) conducts an experiment that was designed to mimic international carbon trading. In his experiment, he included one large buyer, which represented the United States. One differentiating characteristic of this experiment is that the participants in the experiment were given no chance to gain experience with the market by repeating the experiment, a characteristic that Carlén argues comes closer to actual experience in international permit trading, given its novelty. In this experiment, he does not find evidence for distortions through potential market power.

Bohm and Carlén (1999) examine how important market power might be in a related policy setting—the joint implementation component of the Kyoto Protocol. Under joint implementation, investors negotiate to fund specific projects in another country, counting the resulting emissions reductions against the investor's Kyoto obligations. Since joint implementation projects necessarily involve negotiations among a few participants, the possibility of market power is present.

This study does find some evidence of market power, but its influence was small. In general, they found that the negotiations resulted in trades that were close to efficiency and that market power did not prevent successful trades from taking place.

It appears from this review that the evidence from experimental economics is mixed. While one study (Godby 2000) finds that market power could be a sufficiently large problem as to outweigh the other advantages of tradable permits, most studies find market power would not be much of a problem.

## Mechanisms for Controlling Market Power

For any specific traditional or subsidy auction that seems particularly vulnerable to price manipulation, special auctions can be designed to counter this power.[9] Known as incentive-compatible (or Vickery) auctions, they eliminate any incentive for a source to unilaterally use its own bids to control price.

The procedures of an incentive-compatible auction are simple enough. As in a regular auction, each source submits to the auctioneer a demand curve for permits, listing the number of permits desired at each of a number of possible permit prices. The auctioneer sums these bids, chooses the market-clearing price, and awards the permits to those bidding at least as high as the market-clearing price.

So far the procedures are identical to those employed in a conventional auction system. The difference arises in determining the prices to be paid for the acquired permits. In contrast to a single-price auction, where all successful bidders would pay the market-clearing price for all awarded permits, the prices paid for n permits acquired in an incentive-compatible auction by any source would be equal to the n highest rejected bids submitted by other sources.[10] Because these rejected bids are by definition lower than the market price (otherwise they would not have been rejected), incentive-compatible auctions imply lower permit expenditures than more traditional single-price auctions.

This method of price and quantity determination eliminates the price-setting source's incentive to lower the price in an auction by understating its own demand. Although any source understating its demand could acquire fewer permits, it could not lower the price it pays for permits in the absence of collusion with other sources. Because price is determined by the rejected bids of other sources, no source can unilaterally affect the price it pays for permits by artificially raising or lowering its demand. It can only raise its control costs by attempting to influence the process. It is important to note that the auction associated with the U.S. Sulfur Allowance Program is not an incentive-compatible auction (Hausker 1992).

Fortunately, few permit markets contain a large number of direct competitors in the output market. Not only does a typical airshed contain a number of different sources, but in many permit markets the industrial sources emitting a particular pollutant in a given area rarely have much overlap in product markets.

The absence of overlap suggests that in most product markets the permit market would be a relatively (or completely) inefficient vehicle for a predatory source to use to inflict harm on competitors. Many of its competitors in the permit market would not be competitors in the product market. Denying permits to those few rivals would raise its financial burden but would not provide the predatory source with much in the way of commensurate gain.

Some concern has been voiced about the ability of existing firms to use emissions trading as a barrier to entry for new firms. Even though no evidence of this behavior has materialized, the desire to reassure those with these concerns has caused some to create a "set aside," a pool of permits that is available to new entrants.

In the EU ETS, for example, the EU Commission only asks Member States to describe how new entrants can gain access to emissions allowances. No rules dictate whether or not new entrants should be allocated free allowances. Still, all member states guarantee that a certain volume of allowances will be available to new entrants at no cost by creating a set-aside of allowances reserved specifically for new entrants. Allowances from these reserves usually are provided on a first-come, first-served basis (Åhman et al. 2005).

In the Sulfur Allowance Program, the set-aside allowances are available at a predetermined price. This pool of allowances has never been accessed but that doesn't mean that it wasn't useful. It may have increased the political feasibility of the program by providing reassurance to potential entrants and by limiting the potential gains from predatory behavior.

Finally, if concerns about market power arise in permit markets, it is possible to set limits on the concentration of permits that would be permissible for any single source to hold. Though market power concerns in air permit markets have not risen to a level where concentration limits have been imposed, markets for harvesting permits in fisheries have imposed them with some success (National Research Council 1999).

## Programmatic Design Features that Affect Market Power

Though in general market power may not be a serious concern, it is true that some program designs make market power more likely than it would be in less restrictive alternatives. Elements designed to achieve other objectives may have the unfortunate side effect of increasing the vulnerability of emissions trading to market power. Although it is not possible to quantify definitively the seriousness of this problem due to the lack of empirical evidence, it is possible to use the foregoing analysis to identify potential problem areas and to assess their significance at least qualitatively.

## The Baseline Allocation of Control Responsibility

As noted above in the discussion of the Hahn and successor models, the baseline allocation of control responsibility can affect market power. The fact that most permits are distributed free-of-charge means that some specific distribution rules could cause market power potentially to be a more serious problem than would be the case with auctions. On the other hand, it is not obvious that commonly used distribution rules create the kind of situation that is conducive either to price manipulation or competition-reducing market power.

Whereas traditional or subsidy auctions place all sources on the same side of the market, in a grandfathered approach some sources are buyers and some are sellers. Grandfathering can split the participants into buyers and sellers. Depending on the initial distribution rule, a few sources could comprise a significant proportion of the buyers or the sellers, a condition conducive both to price manipulation and competition reduction.

As illustrated in the analysis above, the distribution rule that creates the most problems allocates a disproportionate share of permits to a few large sources. Because of economies of scale, these sources could sell permits without incurring large increases in control costs. The purchasing sources, facing a large deficit of permits and very high marginal control costs, are vulnerable to price manipulation and to any predatory source seeking to put them out of business.

In general, commonly used distribution rules are beneficial in protecting sources from predators. Since the initial allocation generally is economically feasible, existing sources typically would not be forced out of business even if no other source proved willing to sell them any permits. In contrast, failure to acquire any permits in an auction market would, in most cases, mean the closure of the plant.

In existing permit markets, one specific initial allocation does raise market power concerns—the assigned amount allocations in the Kyoto Protocol. In this case, the initial allocation distribution rule was based on 1990 emissions. This choice of a baseline, when coupled with a serious post–1990 decline in economic productivity in Russia and Ukraine, could make them the main sellers of emissions rights to other Annex I countries. Furthermore, due to these surplus permits and the lack of any constraining influence from marginal control costs (since the surplus required no further reductions), the possible effects of monopolistic behavior by Russia and Ukraine in permits trade could reduce the efficiency of permit markets. Preliminary estimates made with the Organisation for Economic Co-operation and Development GREEN model suggest that the influence of market power in this case could reduce the efficiency of the market by about one-third (Burniaux 1999).

As pointed out by Bernard et al. (2003), however, even in this case the Russian decision to exercise market power is not straightforward. Since it is also a major fossil fuel exporter, fossil energy prices are depressed further (due to the lack of permits to justify the associated emissions) to the extent that Russia withholds permits from the permit market, and the value of its exports of energy are reduced. Thus, Russia faces a tradeoff between maximizing its permit revenue and its revenue from fossil energy exports, a fact that constrains its ability to use its potential market power.

## Restrictions on Trading

Spatial concerns about the potential for hot spots have resulted in design innovations that could exacerbate market power concerns. In the bubble policy, for example, offset ratios greater than 1 commonly were required for trade between nonproximate sources. In addition to raising compliance costs by reducing trading opportunities, these rules increase the potential for market power by reducing the number of trading sources.

An alternative approach, adopted in the RECLAIM program, divides the market into separate zones and imposes trading restrictions between zones. Reducing the number of potential trading partners increases the potential for market power. Distant emitters that could serve as an alternative source of permits, thereby limiting the proximate source's ability to raise prices or to harm a competitor, cannot compete in the face of the trading restrictions. Partitioning the market into zones reduces one of the natural checks on market power.

## Treatment of New Sources and Plant Closures

In the U.S. Emissions Trading Program, all new sources in nonattainment areas are required to purchase offsets from existing sources. To the extent these existing sources can exercise market power of either type, the bias against new sources could be made worse. How serious a problem is this market power likely to be?

Since existing sources usually are numerous, it is not obvious that any one of them could exert enough of an influence on the market to affect price or the ability of a competitor to secure permits. The existence of other sources ready to sell permits would limit the ability of any single source to exercise market power of either type.

As more sources move into the area and the degree of control on existing sources is tightened, it would become more difficult to free-up additional permits. Fewer sources would be willing to sell permits because of the high cost of further control. Those sources that do have permits to sell would be in an advantageous bargaining position. The strength of this bargaining position would depend on the vulnerability of new sources and on the marginal cost of further control by the selling source.

The vulnerability of new sources would depend on the availability of alternative production locations. To the extent that locating in this airshed dominates other possible locations, a price-setting existing source would be able to capture some of the location rent associated with this particular desirable location in the permit price. With the availability of other equally suitable locations, however, the existing source would not be able to extract much location rent.

The ability of any existing source to manipulate price also depends on its marginal cost of control. In the circumstance under discussion, the marginal cost of control can be expected to be very high for existing sources. It would be very expensive to make other permits available for sale, thus limiting the capacity of any individual source to influence the market.

There is, however, one major exception to this generally rosy outlook. When a plant shuts down and its permits are banked, the costs associated with creating the permits already would have been incurred; in this case, cost would not act as a brake on the exercise of market power. Given the limited availability of permits from existing sources, this situation could give a single source command over a substantial number of available permits. Banked permits for shutdowns could create the possibility for departing firms to extract location rent from unrelated new sources, particularly if the cap is stringent and few alternative suppliers are in evidence. The most troubling scenario would involve a single source of offsets (generated by a plant closure that has already taken place) facing new sources that are vulnerable in the sense that alternative locations are much more expensive and their entrance into the market is totally dependent on the acquisition of the permits.

Should local control authorities become concerned that a unique circumstance in their trading area has created the threat of market power, the control authority in the United States could use its eminent domain authority to purchase (not confiscate!) shutdown permits by providing just compensation. Just compensation would be defined in terms of previous permit transactions in that area, the rate of inflation, and other relevant factors. The control authority could then resell the permits to some new source at a price that was sufficient to cover its costs (including administration costs). This option rarely would need to be exercised except when the only offsets for sale were the shutdown permits and where the market power threat seemed particularly high. Because it presumably would be exercised only when a willing buyer was available, the cash flow implications for the control authority would be minimal.

# Summary

## *Insights from Theory*

- Two distinct types of market power are possible in transferable permit markets. The first arises when a price-setting source or a collusive coalition of sources seeks to manipulate the price of permits to reduce their financial burden from pollution control. The second stems from the desire of one predatory source or a collusive coalition of sources to leverage market power in the permit market or the output market (or both) for increasing profits in both markets.

- In permit auction (subsidy) markets, price-setting sources purchase (sell) fewer permits than is cost-effective. Control costs are higher than the least-cost allocation, but the financial burdens borne by all sources (not merely the price-setting source) are lower. Air quality is not adversely affected, but revenues received by (paid by) the control authority are reduced (increased).

- The available evidence suggests that price manipulation in permit auctions under the worst circumstances can have a large influence on price, but the typical influence on control costs is rather small. The factors influencing the ability to manipulate demand include the relative size of the price-setting source's demand in relation to the demand of other sources and marginal control costs for both the price-setting and price-taking firms.

- When a grandfathering approach is used instead of an auction or a subsidy market, the degree to which prices can be manipulated depends on the rule used to distribute the baseline control responsibility.

- Market power in the permit market can intensify the degree of noncompliance even if the dominant firm complies. Higher permit prices increase the incentive to cheat by price-takers.

- Market power in the permit market can be used to raise rivals' costs by making permits more expensive and hence to gain an advantage in the product market. Despite this possibility, according to existing models allowing trading still increases welfare.

- The presence of market power can increase the incentive for innovation in abatement equipment, but this effect is smaller for a tradable permits regime than for a regime based upon emissions standards.

## *Insights from Empirical Simulations and Experimental Economics*

- Though the available empirical evidence is very limited, it is remarkably consistent in finding that control costs are not sensitive to market manipulation in air pollution control. Despite finding circumstances where prices and total financial burden were affected dramatically, regional control costs were remarkably insensitive to market manipulation in all ex ante simulations.
- The evidence from experimental economics is mixed. While one study finds that market power could be a sufficiently large problem as to outweigh the

other advantages of tradable permits, most studies find market power is not likely to be much of a problem.

## Implementation Experience

- Actual experience with market power (or its lack) in operating permit markets suggests that the amount of research devoted to market power issues, which is very large, is not a good indicator of the practical importance of this issue. Market power typically has not been a problem in emissions trading.

- Though implementation experience with the use of emissions trading does not uncover any market power concerns in practice, the potential ability of Russia to dominate the emerging Kyoto market for greenhouse gas trading is a possibility worth watching.

- Several design features of emissions trading programs, such as the rule used to distribute permits, restrictions on permit transfers, and the treatment of permits freed up due to plant closures, all could influence the likelihood of market power. Therefore, the potential market power consequences should be considered when those design features are considered.

- Should market power threaten to be a problem in a specific application, remedies such as incentive-compatible auctions, set asides, and concentration limits are available.

## Notes

1. While the Van Egteren and Weber (1996) analysis treats enforcement as exogenous, Stranlund and Dhanda (1999) demonstrate that their results hold up even in the presence of endogenous enforcement. (Endogenous enforcement in this context means that the enforcement authority accounts for the market effects of its monitoring and enforcement strategy and can tailor that strategy to firm characteristics.)

2. Innes et al. (1991) also note that while manipulating the initial endowment to the monopolist can sustain a second-best equilibrium, a uniform tax on emissions cannot. The difference is due to the inability to differentiate with a uniform tax.

3. In Cournot's famous duopoly model, two rivals produce a homogeneous product. Each producer realizes that his rival's quantity decision will impact the price he faces and thus his profits, so he takes the rival's reaction into account. Consequently, each producer chooses a quantity that maximizes his profits subject to the quantity reactions of his rival. Notice that this form of modeling is different from price-setting firms faced with price-taking competitors because in this case the competitors react less passively to the threats. They don't simply take prices—they react strategically.

4. An earlier paper by Malueg (1990) investigated this question and, in contrast, finds that the introduction of trading could reduce total industry profits. According to this model, aggregate profits might fall if the large, low-cost firm reduced its output after trading and the small, high-cost firms increased theirs. Sartzetakis (1997b) shows that cost-minimization in the permit market rules this case out (see footnote 8, p. 80). Both authors find that consumer surplus always rises with trading.

5. Biomass potential is an aggregate index constructed as a weighted sum of biochemical oxygen demand, total nitrogen, and biologically available phosphorus.

6. de Lucia (1974, *86*).

7. From a theoretical point of view, these initial allocations should affect the distribution of payoffs among students but not the size of the total payoff to all students. In a zero-revenue auction, the initial distribution should not affect the potential for market power. The evidence was consistent with this expectation.

8. A double auction involves many sellers and many buyers, as opposed to a conventional auction with one seller and many buyers.

9. The basic work on the properties of these auctions in an emissions trading setting can be found in Lyon (1982, 1986).

10. In another version of this approach, the price paid is equal to the highest unsuccessful bid.

# 8

# Monitoring and Enforcement

Even the most carefully designed regulatory programs can flounder if the enforcement effort is deficient. Establishing emissions trading systems that allocate the control responsibility cost-effectively is of little value if sources regularly fail to comply with the terms of their permits. Ineffective enforcement could undermine the quest for better air quality at lower cost.

The effectiveness of any enforcement program is not only a function of such readily identifiable factors as the size, motivation, and competence of the enforcement staff; the nature of the program makes a difference. Some programs, such as those involving more easily detected violations, are inherently easier to enforce.

Pollution control has certain attributes that increase the difficulty of enforcement. Many of the pollutants are invisible to the naked eye and can be measured only with fairly expensive instrumentation. Although the public at large is victimized by pollution, it is frequently so unaware of the dangers that it cannot be relied upon to assist the regulatory authorities by pointing out violations. With some exceptions, regulatory authorities are on their own.

For the purposes of this book, the main question is how the choice to control air pollution with an emissions trading program affects enforcement and vice versa. Does emissions trading make enforcement easier or more difficult? How have enforcement considerations helped to shape program design? To what extent do the enforcement properties of the emissions trading program reinforce or limit the ability of the program to accomplish its objectives?

## The Nature of the Domestic Enforcement Process

Enforcement of emissions trading generally involves four steps: (1) detecting the violation; (2) notifying the source; (3) negotiating a compliance schedule;

and (4) applying sanctions for noncompliance when appropriate. These responsibilities can be shared among various levels of government. In the United States, the states are assumed to have the primary responsibility for enforcement, but EPA has the authority to step in and bring enforcement actions against polluters failing to comply with their SIP permits.[1] As a matter of practice, EPA takes cases that are particularly complex, that the state has failed to resolve, or that the state has avoided for political reasons.[2]

## Detecting the Violation

Two distinct types of compliance have to be verified: (1) initial; and (2) continuous compliance. The former involves determining that the plant is in compliance when it commences operation of its control equipment. The latter involves verifying compliance during the continuous normal operation of the plant throughout its useful life. While a single set of tests conducted at installation suffices to determine whether the source is initially in compliance, it is inherently more difficult to verify continuous compliance.

Several means are used to detect violations, including self-certification by sources, on-site inspections, and direct monitoring of pollutant flows. Self-certification involves reports from the source as to whether or not it is in compliance. These are based on emissions levels calculated from process control or fuel data. Continuous compliance is ascertained by annual updates to the initial report submitted when the source applied for its operating permit. Though self-certification is the cheapest means of determining compliance, it also may be the least reliable. Sources have an incentive to place their own compliance behavior in the best possible light.

A somewhat more reliable method involves sending trained control personnel on an inspection visit to the emissions site.[3] The inspectors walk through the plant checking the operation of the control equipment, taking readings, and sampling the fuel as appropriate. Certain types of equipment malfunctions or changes in fuel chemical composition can be easily discovered during these inspections but others cannot. Normal, current practice involves announcing the inspections before they take place, precluding random sampling of the actual operating experience.

The degree of scrutiny can be tailored to the likelihood of noncompliance. Even very simple inspections have been found to have a significant deterrent effect if they succeed in identifying potential violations (Wasserman et al. 1992). Therefore, sources that are thought likely to be in compliance may be subjected to less intrusive (and less expensive) inspections. More intensive inspections can be targeted at sources likely to be out of compliance.

Targeting can take advantage of past compliance history. For example, it is possible to start with a less expensive inspection, shifting to more intense scrutiny only if noncompliance is uncovered. As Wasserman et al. (1992) point out, this approach shifts some of the burden of data gathering to the source and postpones resource-intensive inspections until lower-level inspection and monitoring warrant the expense.

## Notifying the Source

Notifying the source of a violation initiates the process of regaining compliance. It serves the twin purposes of providing a beginning date for any legal procedures and stimulating the source to recognize and to deal with the problem. Upon receiving the notification, the source can unilaterally take steps to reduce emissions sufficiently to return to compliance; it can enter into negotiations with the control authority to define a mutually agreeable response; it can question the validity of the noncompliance finding; or it can simply refuse to comply.

## Negotiating Compliance

Unless the source voluntarily submits an acceptable plan showing how it would attain or return to compliance, negotiations are undertaken. These normally attempt to draft a schedule and a means by which the source will attain compliance. Points for negotiation include interim and final deadlines, the types of activities to be undertaken, the kinds of equipment involved, provisions for testing to ensure that compliance has been achieved by the deadlines, and the types of penalties to be levied.

## Enforcing Compliance Orders

If the source violates a compliance order, control authorities usually can take further legal action using an administrative or a judicial response to directly force compliance with the order. Whereas administrative responses are handled internally, judicial responses require taking action in the courts. Though the administrative processes may be similar to those provided by the court system, two advantages of administrative enforcement are that it does not require coordination with a separate judicial agency, and the administrative organization's own administrative law judges usually have more expertise in the matters brought before the court. Therefore, administrative actions usually are resolved more quickly and require less time and expense than judicial actions (Wasserman et al. 1992). However, administrative orders are not self-enforcing. If the order is not complied with, further enforcement action will need to be pursued through the judicial system.

In the United States, suits can be filed either by EPA or by the state involved. Suits filed by EPA face a significant bureaucratic hurdle before they even reach the court—the Department of Justice and the local U.S. attorney must be convinced to pursue the case. At the least, this step adds significant delay.

Further delay ensues once the case reaches the court. It takes time for all the pretrial motions to be filed, for the case to be heard, and for a verdict to be rendered. If the court agrees with the control authority, it can issue a regulatory injunction that sets out detailed compliance schedules and can levy financial penalties for past violations.

### Sanctions

Formal enforcement mechanisms are backed by the force of law and are accompanied by procedural requirements to protect the rights of the individual. These procedures can be either civil or criminal.

Criminal penalties usually are reserved for those situations when a person or facility has knowingly and willfully violated the law or has otherwise committed a violation for which society has chosen to impose the most serious legal sanctions available. One relevant example for emissions trading programs is the use of criminal sanctions when self-reported emissions records have been falsified. Criminal sanctions may include monetary penalties, or imprisonment, or both.

Civil action sanctions may be either administrative (i.e., directly imposed by the enforcement program) or judicial (i.e., imposed by a court or other judicial authority). Field citations, administrative orders issued by inspectors in the field, provide one example of an administrative sanction. Typically, they require the non-complying firm to correct a clear-cut violation and pay a small monetary fine.

Civil sanctions also may be imposed by the courts. When administrative enforcement mechanisms are available, civil judicial responses typically are reserved for use against more serious or recalcitrant violators, when precedents are needed, or where prompt action is important to shut down an operation or stop an activity (Wasserman et al. 1992).

## The Nature of the International Enforcement Process

Since emissions trading moved into the international arena with the Kyoto Protocol coming into force, it makes sense to consider how this institutional context affects both the design and implementation of monitoring and enforcement. To start with the most obvious point, most monitoring and enforcement issues are, of necessity, quite different at the international level. While nations have some legal control over polluting activities in their jurisdiction, international organizations have no similar power. In addition, participation by international agreements by nations is largely voluntary. Hence, draconian sanctions are likely to be counterproductive.

Recognizing these differences has led a number of commentators to search for mechanisms to enhance the self-enforcing capacity of international agreements, a strategy that uses both transfers among participants and aspects of agreement design to make continued participation attractive to member nations.[4]

## Monitoring

Due to the absence of both infrastructure and legal power at the international level, the nations that host the individual pollution sources necessarily would have the major responsibility for monitoring and enforcing the activities of those sources. Ample precedent for this approach has been established by previous international agreements.[5]

Self-reporting of emissions levels by sources to national enforcement agencies is the key to this system, supplemented by cross-checks provided by other sources of data (e.g., fuel consumption data). A standard practice in environmental regulation, self-reporting has the virtue that it is relatively inexpensive, and can provide remarkably accurate information when it is backed by the appropriate sanctions for misrepresentation.[6]

Private monitoring also has become an important check of the self-reporting system. Many bodies of water in the United States, for example, are now protected and monitored by private associations of citizens concerned about that particular body of water. One increasingly typical activity of these voluntary, privately financed associations is hiring a full time "riverkeeper" or "harborkeeper" to continuously monitor polluting activities on that body of water and to monitor the water quality itself. Successful legal actions brought against municipalities and other polluters along the Hudson River in New York state not only have been a powerful force for cleaning up the river but have established some institutional and legal precedents that are now being followed in other parts of the country.

Computer systems and software now have advanced to the point where it is relatively easy to develop automated systems to aggregate and integrate all of these reports, to compare them to alternative sources of data for purposes of validation, and to identify and prioritize suspected cases of noncompliance for action.[7] Both the requisite hardware and software easily could be transferred to nations that sought them.

The direct measurement of emissions is expensive, but fortunately in the case of carbon, is also unnecessary. Because of the close link between the carbon content of the fuel and the amount of carbon dioxide emissions, monitoring the flow of fuels is not a bad proxy for monitoring emissions.[8] This is an important point because monitoring the flow of fuels is much easier than monitoring emissions. While recordkeeping systems already are in place to keep track of the flow of carbon-based fuels across international borders, no counterpart exists for emissions.

How would these monitoring and enforcement activities be financed? As is

already the case in many applications of tradable permits in fisheries management, private entitles could pay a fee to the government for each entitlement
received. This revenue could be used to finance the recordkeeping and surveil-
lance necessary to run the system. Even fees that are too low to have any
significant incentive effects have the potential to generate considerable revenue.

A smaller, but important, monitoring role has been reserved for a newly
designated international monitoring authority. One department in this organization receives the individual country reports and validates them in so far as
possible by cross-checking them with independent sources. Information on
country-by-country compliance is available on the Web.

### Enforcement

What works at the national level would not work or even be necessary in the international context. Oversight of national enforcers requires a different approach.

Compliance with the norms of international law by the nation–states subject to those laws is driven less by the narrow economic motivations that tend
to drive profit-making firms than by the desire to be a part of maintaining a stable, dynamic equilibrium in international relations. This basic tendency can be
reinforced by an appropriately designed global warming agreement; rules universally perceived as fair are more likely to be obeyed than rules perceived as
unfair.

It can also be reinforced by instituting what Chayes and Chayes (1993,
*290–304*) call the principles of accountability and transparency. According to
these principles, governments should be held accountable for their behavior by
rendering their performance transparent to scrutiny by the international community. Requiring the submission of and making public the annual reports to
the international monitoring authority provide the foundation for applying
these principles.

The process of putting in place an international regime for enforcement is
likely to be an evolutionary process. The first steps will be hesitant and will no
doubt prove frustrating for those who seek to transfer the traditional national
model of sanctions into an international setting. Nonetheless, experience shows
that transparency and accountability can be powerful, foundational enforcement principles. Should further action become necessary, these initial steps
can pave the way for a further strengthening of the procedures.

## The Economics of Enforcement

This brief description of the enforcement process is probably sufficient to convey one essential point: it could be very misleading to assume perfect
enforcement when comparing regulatory approaches.[9] Not only will some vio-

lations inevitably go undetected, but not all detected violations result in compliance. At the very least, compliance will be achieved only after a (sometimes substantial) delay.

The degree to which source compliance can be ensured affects, and is affected by, the form of the regulatory policy. Policies that may be cost effective with perfect enforceability, but difficult to enforce, may turn out to be less desirable than easier-to-enforce, but less cost-effective policies. To facilitate the understanding of the effect of the emissions trading program on the costs of air quality control, we begin by considering how imperfect enforcement would affect the behavior of control authorities and sources.

## Cost-Minimizing Source Behavior

A source seeking to minimize costs in an imperfectly enforced regulatory environment must weigh the costs of complying against the costs of not complying. The source is faced with two distinct pathways to achieving compliance, each with its own set of costs:

- The first (and more familiar) category includes the costs of meeting a fixed, predetermined standard, such as expenditures on control equipment, operating costs associated with running the equipment and maintaining it, and the costs of monitoring emissions to verify compliance.

- The second category includes expenditures to achieve compliance by relaxing the standard. These may include lobbying expenditures to amend the regulations or to obtain a variance, as well as litigation expenditures to gain an exemption from, or at least a delay in, compliance.

## The Noncompliance Decision

By formalizing how the cost-minimizing source would compare these alternatives, it is possible to create a picture of source decisionmaking that is helpful in tracing how various elements of the enforcement process affect incentives to comply. Early studies of the economics of enforcement (Downing and Watson 1975; Harford 1978; Malik 1990; Keeler 1991) show that the cost-minimizing source faced with imperfect enforcement would choose a level of compliance at which the marginal cost of compliance was equal to the expected marginal costs of noncompliance. The expected marginal costs of noncompliance are defined as the likelihood that a violation will be detected and a sanction imposed, multiplied by the marginal sanctions once noncompliance has been established (including fines, legal expenses, etc.).

In marked contrast to the effects of noncompliance in regimes controlled by emissions standards, Malik's analysis shows that the principal effect of noncompliance in emissions trading is to alter the equilibrium permit price. Where the number of firms is small, noncompliance on the part of even one firm in the market may affect the pollutant discharge levels of the other firms through its impact on the equilibrium permit price. By lowering the demand for permits, noncompliance lowers the permit price. Lower permit prices means fewer emissions controlled by the other firms.

Another implication of these simple models of regulated source behavior is the overriding importance of the difference between the marginal cost of initial compliance and the marginal cost of continuous compliance. As the standards imposed on a source become more stringent, the marginal cost of initial compliance rises, causing expenditures designed to relax the standards to become more attractive. As the marginal cost of continuous compliance rises, the attractiveness of noncompliance as a conscious strategy increases.

With high initial compliance costs, the amount to be saved by avoiding or delaying compliance increases. Delay becomes more attractive with high compliance costs because the interest earned on the uncommitted funds is larger. These considerations not only suggest that the political pressure on control authorities to relax standards increases as the stringency of the standards increases but that the expected degree of noncompliance rises as well.[10]

Once initial compliance has been demonstrated, subsequent noncompliance need not be as devious as shutting off or by-passing the control equipment, though such actions are not out of the realm of possibility. Noncompliance can also result from a decision to choose a less reliable (and presumably less expensive) control technology or a decision to spend somewhat less on maintenance or upkeep of the equipment than required for continuous compliance.

Downing and Watson (1975) provide an illustrative historical example of how such a bias in control technology selection could occur with lax enforcement. Two different precipitator technologies could be used to meet the new source performance standards for coal-fired power plants. The first, which they label the flexible technology, uses electronic instrumentation to optimize filtering capacity as discharge electrodes fail over the operating cycle. This optimizing capability provides a means of compensating for any deterioration of performance with use so that the degree of compliance can be maintained. The second, a more inflexible technology, provides no such hedge. Though the inflexible technology is cheaper to run, it is also less reliable over the long run.

Simulating the choices of managers of these plants given less-than-perfect enforcement, Downing and Watson (1975) find that the source costs are minimized when the inflexible technology is chosen. Though from society's point of view the flexible technology is preferred, from the point of view of the source, the increased reliability of the flexible precipitator is not worth the cost as long as enforcement is lax.

Sources may prefer inflexible technologies in another sense as well. Typically, inflexible technologies have sharply escalating marginal costs as the technology is operated much above its design capacity. Once it is installed, it is very expensive to increase the degree of control with this technology. Though this inflexibility makes the achievement of the standard much more difficult when more control is needed, this inflexibility can be an attractive feature to sources when they plead their case for relaxed standards. The very high cost of further control makes it easier to convince the courts that further reductions would be economically infeasible and, therefore, unjustified. With lax enforcement, what is cheapest for the firm is not cheapest for the nation.

## *Control Authority Responses*

The control authority, of course, is not powerless in its efforts to ensure compliance because it can raise the cost of noncompliance. Specifically, it can raise the expected cost of not complying by manipulating the two main elements that make up that cost: (1) the likelihood that violations will be detected and sanctions levied; and (2) the level of the sanctions.

In principle, how does the control authority decide how to determine the appropriate level of the expected noncompliance cost? Based on the work of Becker (1968), one method is to set the expected noncompliance cost equal to the expected damage caused by noncompliance. This approach ensures that the source must compensate for, and therefore consider, the harm it causes each time it violates the terms of its permit.

For a given amount of damage caused by noncompliance, this formulation not only requires that higher sanctions be imposed whenever detection probabilities are low, it also requires that payments exceed the damage actually caused.[11] The difference between the value of the damage caused and the size of the monetary sanction is designed to compensate for detection and conviction probabilities that are less than one and to serve as a warning that noncompliance is expensive, even when detection probabilities are low.

This formulation also implies that the expected cost of noncompliance should increase with its duration and intensity. Greater harm triggers larger sanctions. To entertain other designs would undermine the incentive properties of enforcement sanctions. For example, levying the same penalty regardless of the harm caused on all non-complying sources, say $10,000, would not, in general, lead to effective enforcement because it would fail to distinguish between slight and gross noncompliance. Once the source had crossed the line into noncompliance, whether intentionally or not, with this penalty structure, it would not reduce its cost by keeping the degree of noncompliance small. In the language of economics, the marginal (or incremental) deterrence would be zero for fixed penalties that fail to consider the intensity or duration of noncompliance.

One of the intentional, but nonetheless controversial, characteristics of the Becker approach is that it provides support for the notion of "optimal noncompliance." This means that for some occasions, noncompliance is the expected, even desired, outcome. When the cost of compliance is especially high, such as during an equipment malfunction, or the damage is especially low, such as during periods when the air quality exceeds the standards, the source is expected to violate the standard. This outcome is cheaper, not only for the source, but for society as a whole because the cost of coming into compliance exceeds the harm caused by noncompliance. Because the source pays the penalty, the malfunction would be corrected, but extraordinary efforts to maintain compliance while the repairs were being made only would be undertaken if they were less expensive than the harm caused.

Even ardent supporters of this philosophy recognize that it has a serious flaw. Reliable estimates of the harm caused by noncompliance are very difficult to acquire. The absence of these estimates undermines the basis for establishing the level of the sanctions in the Becker-type approach.

## An Alternative Philosophy of Compliance

An alternative approach, having the substantial virtue that it rests on observable (and therefore calculable) factors, includes any cost savings the source might have achieved by noncompliance in the penalty. The objective of this approach is zero noncompliance rather than optimal noncompliance. It seeks to eliminate noncompliance altogether by removing any cost advantages the source would accrue from noncompliance.

Actually, this approach does not succeed in creating incentives for complete compliance since it does not take the likelihood of detection into account. In principle, the noncompliance penalty should be equal to the cost savings from noncompliance divided by the probability that a non-complying source would be caught and sanctions imposed. Only by taking the actual probability into account would the expected cost of noncompliance be equal to the expected cost savings any source could expect to accrue from noncompliance.

In practice, rather than base the penalty on weakly supported estimates of the likelihood of sanctions being imposed in each instance of noncompliance, the noncompliance penalty authorized by U.S. statutes makes no correction for the likelihood of detection.[12] Since this approach could be expected to completely deter noncompliance only if all violations were detected and sanctioned (by no means a certain occurrence), this system can be expected to result in more noncompliance than a system involving a correctly defined penalty. It cannot yet be determined with any accuracy how serious a problem this is in practice.

Since this approach does not guarantee complete compliance, what would control authorities have to do to achieve complete compliance? Stranlund, Chavez, and Field (2002) examine this question in the context of a simple model that allows them to derive some remarkably useful and intuitive results.[13]

For complete compliance in competitive markets the authorities would need to assure that two conditions hold

$$(1)\ p < \pi(f + g);\ \text{and}\ (2)\ p < f$$

where

$p$ = market price of permits;

$\pi$ = probability that a source will get audited, which is assumed to be sufficient to discover a violation if one exists;

$f$ = the per unit fine levied for emissions violations; and

$g$ = the per unit fine for under-reported emissions.

These two conditions not only provide firms with the proper incentive to submit truthful reports of their emissions, since the expected cost of lying exceeds the cost of acquiring additional permits, but it also guarantees that each firm will hold enough permits to cover their emissions. Given the incentive to submit truthful reports (guaranteed by the first condition), the firms would always choose to hold enough permits if the price of acquiring those permit is less than or equal to the per-unit fine for not having them.

## *Allocating Enforcement Resources*

How should a budget-constrained enforcement authority distribute its enforcement effort among heterogeneous, noncompliant firms in a transferable emissions permit system? In modeling enforcement agency choices, Stranlund and Dhanda (1999) find, somewhat surprisingly, that a uniform (as opposed to targeted) monitoring and enforcement strategy minimizes aggregate noncompliance given the monitoring and enforcement budget.

If the firm behavior behind their model turns out to be empirically valid, these results imply that regulators accrue no advantage from applying more intense monitoring and enforcement efforts to firms that employ less-advanced emissions control technologies, or that use dirtier production processes, or that differ in any other fundamental way due to the equilibrating nature of the permit market.

The logic behind this finding is straightforward. Each firm chooses its emissions so that its marginal abatement cost is equal to the permit price, and it chooses the number of permits to hold so that its marginal expected penalty also is equal to the permit price. Since all firms face the same permit price, marginal abatement costs and marginal expected penalties are equal to each other and equal across firms. This implies that all adjustments to increased monitoring and enforcement are in the form of acquiring or selling more permits, not in choosing a different level of emissions.

## Instrument Choice

Can we say anything about how imperfect enforcement affects the relative desirability of emissions taxes and emissions trading? Montero (2002) finds that when cost and benefit uncertainty is combined with incomplete enforcement, a quantity instrument generally outperforms a price instrument. In fact, he finds that if the slopes of the marginal benefit and marginal cost curves are the same, the quantity instrument would be preferred.

The reason for this result is that in a quantity regime with incomplete enforcement, the effective (or observed) amount of control is no longer fixed but rather endogenously determined by the actual (ex-post) cost of control. Indeed, if the marginal cost curve proves to be higher than expected by the regulator, more firms would choose not to comply, and consequently, both the effective amount of control and the cost of control would be lower than expected. In effect, incomplete enforcement provides a kind of safety valve in a quantity regime that is not available in a tax regime. When costs prove to be higher than expected, some firms choose not to comply, increasing the effective amount of pollution but lowering the cost consequences of the error.

How does the noncompliance incentive differ between emissions trading and uniform emissions standards? Keeler (1991) finds that emissions trading could result in more pollution than a system of uniform standards when the enforcement of emissions levels is incomplete. Whether the emissions trading noncompliance incentive is higher depends on the nature of the penalty function. If the market-clearing permit price were less than the expected penalty for the initial unit of violation, then all firms would comply and emissions-trading emissions would be less than under emission standards. When the market-clearing price rises above the marginal expected penalty, however, all firms in the market would minimize expected costs by exceeding authorized emissions levels.

Characterizing this relationship has important implications. First, it suggests that for a given noncompliance penalty, emissions trading would be more likely to encourage noncompliance in permit markets characterized by high permit prices. Second, since the marginal expected penalty depends on the likelihood of getting caught, detection probabilities (and the monitoring efforts that underlie them) are important determinants of the demand for permits.

# Current Enforcement Practice

## Monitoring Emissions

How did the introduction of emissions trading affect the enforceability of the air pollution control program? This is not an easy question to answer because in some ways it made enforcement easier, while in others it made it more difficult.

**The Lower Compliance Cost Effect.** The most positive effect on enforcement comes through the ability of emissions trading to reduce compliance costs. Reduced compliance costs not only mean less incentive for sources to seek relaxation of the standards or to entertain noncompliance as an intentional strategy, they also mean more incentive for the control authorities or the courts to enforce the requirements vigorously.

Because of the high costs of compliance associated with the command-and-control system, it paid sources to invest in relaxing the standards or delaying compliance as long as possible. Several examples of successful use of these tactics include statutory exemptions, amendments to state regulations, and variances granted either by the control authorities or the courts. Because investments in delaying and avoiding compliance are themselves expensive, they are only justified when the benefits received are high. When compliance costs are lowered, as they are in an emissions trading program, the benefits from this type of investment (avoided or delayed compliance costs) are diminished. Complying with the regulations becomes relatively more attractive.

The incentives for enforcers to ensure compliance are similarly bolstered by lower compliance costs. Control authorities and courts are understandably reluctant to force a source to install very expensive equipment when the cost seems out of line with the accrued benefits. As the simulation models described in Chapter 3 make clear, the command-and-control policy distributes the burden in such a way that some firms bear a disproportionately high cost. When these sources bring forth their appeals, they have a stronger case for special treatment than they would in the presence of an emissions trading program. By providing lower cost, alternative ways of attaining compliance (namely, by acquiring permits), the emissions trading program makes compliance easier to enforce. The consequences of enforcing the law are less severe than they would be under the command-and-control program.

**The Financing Effect.** The financing of continuous emissions monitoring (CEM) systems by the regulated utilities is indicative of a larger tendency in other tradable permit markets, most notably fisheries (National Research Council 1999). In tradable harvesting quota systems in fisheries, it is becoming more common to finance monitoring and enforcement activities from the revenue flowing from a small tax on quota. Transferring the financial burden of monitoring and enforcement to the source that causes the need for it seems to have a certain political resonance. Since emissions trading usually can reduce costs by a great deal, using part of those gains to finance monitoring and enforcement seems to be an idea whose time has come.

Not all of the aspects of an emissions trading program facilitate enforcement, however, as the following cases illustrate.

**Using Unreliable Emissions Estimates**. Determining compliance in emissions trading requires comparing actual emissions with authorized emissions. If actual emissions are estimated imperfectly, the enforceability of emissions trading is diminished.

While modeling and monitoring both may be useful for determining equivalency, they are not equally useful for determining continuous compliance. Because modeling generates a one-time estimate accomplished before the fact, it can address only the question of initial compliance. The estimate also may be a very poor indicator of actual emissions.

To some extent, existing programs have been specifically designed to eliminate this monitoring concern. The Sulfur Allowance Program, for example, requires utilities to install and to pay for CEM, which is the continuous measurement of pollutants emitted into the atmosphere in exhaust. EPA must certify the CEM system before it can be used in the Sulfur Allowance Program. Continuous monitoring makes it easier to verify continuous compliance. To the extent that continuous monitoring is chosen by the source at its expense, current programs eliminate the monitoring bias.

Not all sources in all programs use or can use continuous monitoring, even in industrialized countries. In the RECLAIM program, major sources have to install CEM systems, but in smaller facilities, cheaper but less accurate systems based upon estimated emissions are allowed. Audits of these reports are supplemented by on-site inspections.

As shown theoretically in Montero (2005), in some cases the optimal policy design in the presence of imperfectly monitored emissions is to use permits in some markets and to combine permits and standards in others. In his model, it is highly unlikely that the optimal policy would be to use standards alone. In any case, the optimal instrument design and choice will ultimately depend on the cost structure of the group of affected sources.

That model involves a trade-off between cost-savings and possible higher emissions when comparing permits and standards under incomplete monitoring. On one hand, the permits policy retains the well-known cost-effectiveness property of conventional permits programs (i.e., those based on actual emissions); permit trading allows heterogeneous firms to reduce their abatement and production costs. On the other hand, the permits policy can sometimes provide firms with incentives to choose combinations of output and abatement technology that may lead to higher aggregate emissions than under standards.[14] Thus, this model suggests that the permits policy is likely to work better when abatement and production cost heterogeneity across firms is large. In contrast, as heterogeneity disappears, the advantage of permits is reduced, and standards might work better provided that they lead to lower emissions. For the special case of Santiago, Chile, the cost structure of the group of affected sources was such that in retrospect the permits policy seems to have been the right choice.[15]

**The Banking Effect.** As demonstrated in Chapter 5, banking clearly helps in both lowering costs and in stabilizing prices in the permit market. Stability in prices in emissions markets is important for a variety of reasons, including encouraging the appropriate level of investment in abatement technology. Banking can, however, lead to lower rates of compliance, since benefits from underreporting emissions are greater when unused permits can be banked for future use or sale.

To see whether the influence of banking on compliance would show up in a laboratory setting, Cason and Gangadharan (2005) conduct an experiment in which emissions trading participants are subjected to emissions uncertainty. Specifically, the subjects faced exogenous, random, positive, or negative shocks to their emissions levels after they had made production and emissions control plans. In some sessions, subjects were allowed to bank their unused permits for future use.

During a reconciliation period, permits could be bought or sold to respond to the emissions shock. Emissions were then reported to the regulatory authority and participants were placed in groups with different inspection rates depending on their compliance history. The authors of this study find that the hypothesis that banking does not affect compliance rates could be rejected by the experiments. On average, actual emissions exceeded allowed emissions in every session where banking was allowed, but in only one-half of the sessions without banking.

In terms of actual experience, Wasserman et al. (1992) suggest that the pattern of violations recorded in the Lead Phase-out Program was consistent with the belief that the introduction of banking initially intensified incentives for noncompliance. She also suggested, however, that banking was a crucial aspect of the program and that subsequent audits and penalties brought compliance back in line.

## Tracking Authorized Emissions

Compliance with the rules of an emissions trading system generally requires surrendering sufficient permits during a reconciliation period to cover the amount of actual emissions during that period. For this reconciliation to be reliable, an accurate official record of authorized emissions and the permit holdings that underlie them must be maintained and actual emissions must be known.

Functioning much like a bank, registries typically are automated databases that keep track of all initial holdings and changes to those holdings through transfer or use. Typically, each permit is uniquely identified by a serial number.

The Allowance Tracking System (ATS) in the Sulfur Allowance Program records the holdings not only of regulated utilities—the sources that must comply—but also of others. The ATS contains two different types of accounts: unit accounts and general accounts.

- Unit accounts keep track of permit holdings (and, hence, authorized emissions) for all regulated sources. Unit accounts record allocated allowances, remove allowances surrendered or sold during the reconciliation period, and add allowances acquired from other sources.

- General accounts can be created by anyone to hold or trade allowances. Any individual or group, including a utility, can open a general account by submitting the appropriate form. General accounts can be used by utilities that wish to pool allowances across units, by brokers or investors to store allowances that have been acquired but not yet sold, and even by public interest groups wishing ultimately to retire a portion of the available allowances from the market, thereby denying their use for authorizing emissions.

Are monitoring and recordkeeping burdens increased or reduced under emissions trading compared to a command-and-control approach? According to an EPA analysis of the RECLAIM program (U.S. EPA 2002), they are increased because more information is required. The shift from a technology and rate-based program to a mass/trading-based program changes the nature of the required information.

Regulating emissions under command-and-control normally only involved mandating a specific control technology and making sure it was installed and running. Under RECLAIM, most facilities do not have equipment-specific control technology regulations; rather, facilities must monitor and report total emissions from all pieces of equipment and processes. The good news is that this system provides much better information that can provide the basis for much better control over total emissions than a command-and-control system but only by committing more resources to monitoring and recordkeeping.

Technology has played an important role in improving the quality and lowering the cost of the data handling involved in effective monitoring and recordkeeping. For example, in the U.S. Sulfur Allowance Program, both the collection and dissemination of the information derived from the continuous emissions monitors is now handled via the Web. Special software has been developed to take individual inputs and generate information both for the public and for EPA enforcement activities. According to Kruger et al. (1999), the development of this technology has increased administrative efficiency, lowered transactions costs, and provided greater environmental accountability.

One unexpected finding that emerges from ex post evaluation is the degree to which the number of errors in pre-existing emissions registries are brought to light by the need to create accurate registries for emissions trading programs (Montero et al. 2002; Pendersen 2003; and Hartridge 2003). Although inadequate inventories plague all quantity-based approaches, emissions trading seems particularly effective at bringing them to light.

## Penalty Structures

**Sulfur Allowance Program.** The penalties for noncompliance in the Sulfur Allowance Program have two components. First, every ton of actual emissions over the level covered by allowances is assessed a penalty. Although this penalty was $2,000 per ton in 1990, it is indexed to inflation. Second, excess tons in one year (the amount by which actual emissions exceed the surrendered allowances) are subtracted from the allocation in the following year. Since the market-clearing price in the 2005 auction was $690/ton, this is considerably lower than the penalty; compliance clearly minimizes costs.

**RECLAIM.** If an audit reveals probable noncompliance for RECLAIM facilities, the company is given an opportunity to review the basis for the finding and to present additional data. For any violation of RECLAIM, the executive officer may seek an administrative penalty of up to $500 for every 1,000 pounds of excess emissions for every day it persists.

During the deregulation crisis, the RECLAIM authority removed existing power plants from the RECLAIM market and required them to install "best available retrofit" air pollution controls. It also established a temporary mitigation fee program, where power plants would pay $7.50 per pound of $NO_x$ emissions in excess of their RECLAIM credit holdings. The South Coast Air Quality Management District used the fees to acquire emissions reductions from other mobile, area, and stationary sources. As in the normal penalty program, excess emissions were deducted from the facility's future account holdings. Once emissions reductions were generated from the mitigation fees, deductions were credited back to the contributing facility's account.

**Lead Phase-out Program.** According to a leading EPA enforcement official (Wasserman et al. 1992), a high degree of voluntary compliance was expected in the Lead Phase-out Program because self-reported outcomes could be corroborated with an outside source of information (manufacturers of lead additives) to verify refiners' reports. In addition, the regulated firms, primarily large refiners, were expected to be quite vulnerable to public opinion.

In contrast to these expectations, the initiation of the audit program late in 1986 revealed substantial noncompliance. The distribution of violations through time demonstrated that while early audits uncovered instances of severe noncompliance, making the result of those audits public tended to deter new violations. Many of the violations detected through audits were large, and the enforcement actions taken against the violators were given wide publicity. Following that publicity, in 1987 new violations declined to about a third of their former level. This pattern suggests that the audits, the penalties, and the transparency of the process successfully reduced new illegal activity through their deterrent effect.

## Promoting Reliable Self-Reporting

Finally, since self-reporting is such an important component of these programs and truthfulness is essential, how do the programs promote truthfulness?

One way is through the use of criminal sanctions for false reporting. How criminal sanctions are employed in the Sulfur Allowance Program illustrates their use. Since the program was authorized under the Clean Air Act, it is covered by that act's authorization of the use of criminal sanctions for false reporting.[16] Importantly, the act also requires each source to identify a single individual as the "designated representative" of the regulated unit. That individual bears the responsibility, on behalf of that firm, for submitting truthful reports. EPA can readily impose criminal liability on this person if a company has submitted false or misleading data. Criminal sanctions under the law include penalties of up to 15 years in jail.

Third-party verification is another strategy to improve the reliability of self-reported estimates of emissions. The EU ETS is launching a different form of monitoring activity in which externally verified self-reporting is being used in the place of continuous emissions monitors, the typical form of monitoring in the United States. This form, which is analogous to existing systems of financial reporting, could prove to be more appropriate for the heterogeneous sources of greenhouse gas emissions, especially those for which CEM is not practical.

## The Bottom Line

The enforcement experience for emissions trading is mixed. On one hand, it has tended to enhance enforcement by lowering compliance costs, by providing more flexibility in how the standards are met, and by allowing these lower costs to be used as leverage for gaining agreement to install continuous emissions monitors. On the other hand, it enlarges the enforcement burden by requiring reliable information on emissions, not merely verifying that the technology is in place or that the maximum emission flow rate has not been exceeded and, in the absence of adequate monitoring of actual emissions and tracking of allowed emissions, provides more opportunities for noncompliance.

The actual enforcement records of emissions trading programs reflect this ambiguity. For some programs like the Sulfur Allowance Program, coupling an effective CEM program with penalties that exceed permit prices has meant complete, or near complete, compliance (Ellerman et al. 2000). Very high compliance rates also have been achieved by the $NO_x$ Budget Trading Program.[17]

Although the Lead Phase-out Program experienced some early increase in noncompliance after the adoption of lead trading, it subsequently demonstrated that coupling an aggressive audit program with transparency of the resulting enforcement actions can restore compliance and act as an effective deterrent to noncompliance (Wasserman et al. 1992).

Though any program based upon fixed quantitative limits on emissions, including emissions trading, can create compliance problems during periods of extreme stringency, such as occurred in California, the RECLAIM experience suggests that the flexibility provided by emissions trading can help to reduce aggregate noncompliance (Harrison 2002).

The RECLAIM experience also raises another important point. Small companies may not be able to offset the additional monitoring and recordkeeping costs with the savings they may accrue through the flexibility of emissions trading in the same way that larger companies can. Successfully integrating small companies into emissions trading programs requires the development of monitoring and enforcement procedures that are both effective and affordable at that scale (U.S. EPA 2002).

# Summary

## *Background*

- Formal enforcement mechanisms may be based upon either criminal or civil procedures. Criminal sanctions usually are reserved for those situations in which a person or facility has knowingly and willfully violated the law, such as submitting a false emissions report, or has otherwise committed a violation for which society has chosen to impose the most serious legal sanctions available. Civil sanctions may be either administrative (i.e., directly imposed by the enforcement program) or judicial (i.e., imposed by a court or other judicial authority). When administrative enforcement mechanisms are available, civil judicial responses typically are reserved for use against more serious or recalcitrant violators, when precedents are needed, or where prompt action is important to shut down an operation or stop an activity.

- Compliance with the rules of an emissions trading system generally requires surrendering sufficient permits during a reconciliation period to cover the amount of actual emissions during that period. For this reconciliation to be reliable, an accurate official record of authorized emissions and the permit holdings that underlie them must be maintained and the measure of actual emissions must be accurate.

## Findings from Conceptual Analysis

- In emissions trading programs where the number of firms is small, noncompliance on the part of even one firm in the market may affect the pollutant discharge levels of the other firms through its impact on the equilibrium permit price. Because a non-complying firm needs fewer permits, noncompliance is likely to lead to a lower permit price, leading other firms to purchase more, thereby offsetting, to some extent, the effects on aggregate noncompliance.

- One of the implications of this price-induced, offsetting reaction is that a uniform, as opposed to targeted, monitoring and enforcement strategy minimizes aggregate noncompliance given the monitoring and enforcement budget. This is different than for emissions standards, where the optimal strategy involves targeting monitoring and enforcement efforts.

- Complete compliance can be achieved by ensuring that: (1) the expected cost of noncompliance (the probability of being detected times the sum of the fines for underreporting emissions and excess emissions) is higher than the permit price; and (2) the fine per unit of excess emissions is higher than the permit price. These two conditions not only provide firms with the proper incentive to submit truthful reports of their emissions (since the expected cost of lying exceeds the cost of acquiring additional permits), but it also guarantees that each firm will hold enough permits to cover their emissions. Given the incentive to submit truthful reports (guaranteed by the first condition), the firms would always choose to hold enough permits if the price of acquiring those permits is less than or equal to the per-unit fine for not having them.

- From society's point of view, investing in control technologies that offer more reliable monitoring, all other things being equal, would be preferred. From the point of view of the source, however, better monitoring may be an externality and, hence, not worth the price as long as enforcement is lax.

- As compared to a case with complete enforcement, emissions trading performs relatively better than a price instrument when incomplete enforcement is coupled with cost and benefit uncertainty. In effect, incomplete enforcement provides a kind of safety valve in a quantity regime that is not available in the absence of trading. When costs prove to be higher than expected, under emissions trading some firms choose not to comply, increasing the effective amount of pollution, but the cost consequences of the error are diminished.

## *Ex Post Analyses of Implementation Experience*

- Because of its focus on ensuring initial compliance with a set of technology-based emissions standards, the command-and-control approach has neglected the need to more effectively monitor and enforce continuous compliance. The evidence is clear that initial compliance does not imply continuous compliance.

- Noncompliance can take several forms, including delays in reaching ultimate compliance, poor maintenance or operation of control equipment, or the willful operation of control processes at less-than-necessary capacity.

- Transferring the financial burden of more effective monitoring and enforcement to the source that causes the need for it seems to have a certain political resonance. Since emissions trading usually can reduce costs by a great deal, using part of those gains to finance monitoring and enforcement though a tax on emissions allowances seems to be an idea whose time has come.

- Technology has played an important role in improving the quality and lowering the cost of the data handling involved in effective monitoring and recordkeeping.

- The shift from a technology and rate-based program to an emissions trading program changes the nature of the required monitoring information. Regulating emissions under command-and-control normally only involved mandating a specific control technology and making sure it was installed and running. Under emissions trading, most facilities do not have equipment-specific control technology regulations; rather, facilities must monitor and report total emissions from all pieces of equipment and processes. In addition, in contrast to command-and-control, emissions trading requires registries to keep track of authorized emissions and changes to authorized emissions caused by trading. Although this enhanced monitoring and tracking system provides much better information that can provide the basis for much better control over total emissions than a command-and-control system, it also costs more. New software developments are reducing those costs over time.

- One apparent effect of the move toward emissions trading has been an improvement in the emissions inventories maintained by control authorities.

- Penalties for excess emissions in modern emissions trading programs typically have two components: (1) a financial penalty on excess emissions that is larger than the permit price; and (2) the subtraction of any source's excess emissions in one commitment period from the source's allocation in the subsequent commitment period.

- In the absence of continuous emissions monitoring, truthfulness in self-reporting can be promoted through specifying a designated representative for each reporting source who is legally responsible for the veracity of the submitted reports and the use of criminal sanctions for false reporting. Third-party verification of reports, such as that currently used for financial reporting, also can help.

- The enforcement experience for emissions trading is mixed. On one hand, it tends to enhance enforcement by lowering compliance costs, by providing more flexibility in how the standards are met, and by using these lower costs as a way to gain agreement to install continuous emissions monitors. On the other hand, it enlarges the enforcement burden by requiring reliable information on emissions, not merely the technology in place or the maximum emission flow rate, and provides more opportunities for noncompliance when registry systems are inadequate or actual emissions are not accurately reported.

- The actual enforcement records of emissions trading programs reflect this ambiguity. For some programs, like the Sulfur Allowance Program and the $NO_x$ Budget Trading Program, coupling an effective continuous emissions monitoring program with penalties that exceed permit prices has meant complete, or near complete, compliance. On the other hand, considerable noncompliance was experienced during the early years of the Lead Phase-out Program.

- As the experience in the Lead Phase-out Program demonstrates, however, in cases where compliance can be a problem, coupling an aggressive audit program with transparency of the resulting enforcement actions can restore compliance and act as an effective deterrent for noncompliance.

- This review also suggests that small companies may not be able to offset the additional monitoring and recordkeeping costs associated with emissions trading through abatement cost reductions in the same way that larger companies can. Successfully integrating small companies into the monitoring and enforcement component of emissions trading programs requires the development of monitoring and enforcement procedures that are both effective and affordable at that scale.

## Notes

1. The federal and state responsibilities are spelled out in 42 U.S.C. 7413. The reporting requirements that form the basis for enforcement actions are found in 42 U.S.C. 7414.

2. Melnick (1983, *197*).

3. Federal and state regulators performed about 17,800 routine inspections each year in fiscal years 1998 and 1999 and found that about 88–89% of the facilities complied with the terms in their permits (U.S. General Accounting Office 2001).

4. For one review of these strategies in the climate change context, see Carraro (2002).

5. Since 1973, for example, the Washington Convention on International Trade in Endangered Species of Wild Fauna and Flora has established worldwide trade controls based on a permit system administered completely by the individual nations. For other examples, see Sand (1991, *259–261*).

6. In the United States, virtually all environmental statutes now make the falsification of environmental compliance reports a criminal offense, with the responsible individual subject to personal fines and imprisonment. For further details, see Segerson and Tietenberg (1992).

7. The nature of the system used to support the United States-lead trading system is described in Nussbaum (1992). Kete (1992) describes in some detail the system currently being developed to support the sulfur oxide trading system that is an integral part of the U.S. approach to reducing acid rain.

8. It is, however, not a perfect proxy. Since combustion is the source of the carbon dioxide emissions, oil used for lubrication does not pose the same problem as oil used as a fuel. As a practical matter, this can be handled by giving production facilities a credit for all oil sold for lubrication.

9. For example, according to the U.S. General Accounting Office (2001), routine inspections do not necessarily identify instances in which the facilities have made physical or operating changes that could increase emissions and require revising their permits. Following up on that concern, EPA found that 76% of the wood product facilities it investigated had made operational changes without revising their permits. Moreover, EPA's investigations in the refinery industry found widespread underreporting of emissions from leaking valves and other equipment.

10. In this model if, as the degree of noncompliance increases, the expected marginal cost of noncompliance rises more slowly than the marginal cost of compliance, it is even possible for actual emissions to increase as the standards are made more stringent. See Harford (1978, p. 33).

11. The sanction would be equal to the damage caused only for those circumstances where the detection probability was 1.

12. 42 U.S.C. 7420.

13. See also Stranlund and Chavez (2000)

14. The fact that emissions trading could lead to higher total emissions than command-and-control also was found earlier by Hahn and Axtell (1995).

15. This information was provided by Juan-Pablo Montero in a personal communication and is based upon econometric work he is about to publish in a forthcoming Oxford University Press book edited by Charles Kolstad and Jody Freeman.

16. Section §113(c)(2)(A).

17. In 2003, of the total affected population of approximately 1,000 units, all but 7 were in compliance (U.S. EPA 2004).

# 9

# Lessons

The preceding chapters have reviewed the conceptual, empirical, and ex post implementation evidence from emissions trading programs. This chapter characterizes the state of the art and the lessons that might be extracted from this review. The lessons have been broken down into four categories: (1) lessons about the effectiveness of the programs; (2) lessons about instrument choice and the design of emissions trading programs; (3) lessons about the ex post evaluation process itself; and finally (4) lessons about the expectations created by theory.

## Lessons about Program Effectiveness

Two types of studies have been used to assess cost savings and air quality impacts of emissions trading programs: ex ante analyses that depend on computer simulations and ex post analyses that examine the actual implementation experience.

A substantial majority, though not all, of the large number of ex ante studies find the command-and-control outcome to be significantly more costly than the least-cost alternative. These studies also find that whenever the need for additional reduction is so severe that the control authority has no choice but to impose emissions standards that are close to the limit of technological feasibility, the immediate potential cost savings typically are very small, though these savings can rise over time as new technologies are introduced.

Ex post cost savings differ from the ex ante estimates of potential savings for three main reasons: (1) the cost of the least-cost allocation may be measured with some error; (2) the cost of the command-and-control allocation may be measured with some error; and (3) the implemented emissions trading program can differ considerably from the idealized programs modeled by the ex ante analyses.

Until recently, it appeared that emissions trading was introduced only after more familiar systems had been tried and proved inadequate. It now appears that the introduction of new emissions trading programs has become easier as some positive outcomes have resulted from implemented programs.

To date, using free distribution of permits, as opposed to auctioning them off, seems to be a key ingredient in the successful implementation of emissions trading programs. The historical record, however, also makes it clear that not every attempt to introduce emissions trading for controlling air pollution has been successful; some attempts to implement emissions trading programs have failed.

Conventional wisdom holds that emissions trading affects costs but not air quality, which seems at best to be an oversimplification. In retrospect, we now know that the feasibility, level, and enforcement of the emissions cap all can be affected by the introduction of emissions trading. In addition, emissions trading may trigger environmental effects from pollutants that are not covered by the limit. While most of these effects are desirable, some are detrimental.

In general, air quality has improved substantially under emissions trading, but with the exception of the Sulfur Allowance Program and the Lead Phaseout Program (where the case seems clear), the degree to which credit for these reductions can be attributed solely to emissions trading, as opposed to exogenous factors or complementary policies, is limited.

In general, the evidence seems to suggest that by lowering compliance costs, tradable permit programs facilitate the setting of more stringent caps. In air pollution control, the lower costs offered by trading were used in initial negotiations to secure more stringent pollution control targets (the Acid Rain, Lead Phase-out, and RECLAIM programs) or earlier deadlines (Lead Phase-out Program).

Pollutants other than those being targeted by the emissions trading program also have been affected by leakage. Leakage occurs when pressure on the regulated resource is diverted to an unregulated, or lesser regulated, resource, as when polluters change their process to emit different pollutants or move their polluting factories to countries with lower environmental standards. In some cases, leakage can intensify the positive effects of a program, as is the case when the control of greenhouse gases results in substantial reductions of other air pollutants associated with the combustion of fossil fuels. But in others, such as when emissions are diverted to unregulated areas, the effects on other resources can be detrimental. Generally, the evidence to date suggests that leakage effects have been small.

The magnitude of the positive air quality effects under credit-type programs, such as the U.S. ETP, has been smaller and the achievements have come more slowly than expected. Credit programs seem to be characterized by more transactions costs and more administrative costs than cap-and-trade programs.

Although hard evidence on the point is scarce, a substantial amount of anecdotal evidence is emerging about how tradable permit programs can change the way environmental risk is treated within polluting firms. This evidence suggests that environmental management use to be relegated to the tail-end of the decisionmaking process. Historically, the environmental risk manager was not involved in the most fundamental decisions about product design, production processes, or selection of inputs. Rather, he was simply confronted with the decisions already made and told to keep the firm out of trouble. This particular organizational assignment of responsibilities inhibits the exploitation of one potentially important avenue of risk reduction: pollution prevention.

Because tradable permits put both a cap and a price on environmental risks, it tends to get corporate financial people involved. Furthermore, as the costs of compliance rise in general, environmental costs become worthy of more general scrutiny. Reducing environmental risk can become an important component of the bottom line. Given its anecdotal nature, the evidence on the extent of organizational changes that might be initiated by tradable permits should be treated more as a hypothesis to be tested than a firm result, but its potential importance is large.

While economic theory treats markets as if they emerge spontaneously and universally as needed, in practice in unfamiliar markets such as these, the participants (both regulators and emitters) frequently require some experience with the program before they fully understand and behave effectively in the market for permits. This finding is potentially important for the implementation of the Kyoto Protocol's Clean Development Mechanism, which involves the creation of transferable credits in developing countries.

Although few detailed ex post studies have been accomplished, completed studies typically find that cost savings from introducing emissions trading are considerable but less than would have been achieved if the final outcome were fully cost-effective.

The literature contains some support for the fact that emissions trading encourages both emissions reducing innovation and the adoption of new, available emissions reducing technologies. It also contains evidence, however, to support the opposite proposition, particularly when cheaper alternatives to new technology, such as fuel switching, exist.

Two distinct types of market power are possible in emissions trading. The first arises when a price-setting source or a collusive coalition of sources seeks to manipulate the price of permits to reduce their financial burden from pollution control. The second stems from the desire of one predatory source or a collusive coalition of sources to leverage market power in either the permit market or the output market, or both, for increasing profits in both markets.

Though implementation experience with the use of emissions trading does not uncover any market power concerns in practice, computer simulations suggest that the potential ability of Russia to dominate the emerging Kyoto market for greenhouse gas trading is worth watching.

One unexpected point that emerges from ex post evaluation of tradable permit systems is the degree to which the number of errors in pre-existing emissions registries are brought to light by the need to create accurate registries for emissions trading. Although inadequate inventories plague all quantity-based approaches, tradable permits seem particularly effective at bringing them to light and getting them corrected.

The case studies (especially for RECLAIM and the Kyoto trading system) reveal some tendency to overallocate quota in the initial years of a program to improve its political feasibility. In the absence of a ratchet mechanism, this may result in a less stringent cap, at least initially.

Emissions trading programs are sometimes thought to be a dangerously rigid approach to resource management. This expectation is based upon the belief that, once instituted, property rights become entrenched and therefore impervious to change. In fact, implemented tradable permit programs such as the ozone-depleting gas program have exhibited a considerable amount of flexibility and evolution over time. In addition, as noted below in the section on adaptive management, new design elements have increased this flexibility.

The compliance incentives for emissions trading are mixed. On one hand, it has enhanced enforcement by lowering compliance costs, by providing more flexibility in how the standards are met, and by using these lower costs as a way to gain agreement to install continuous emissions monitors. On the other hand, it has enlarged the enforcement burden by requiring reliable information on emissions, not merely the technology in place or the maximum emissions flow rate, and, in the absence of adequate registries or adequate measurement of actual emissions, has provided more opportunities for noncompliance.

The shift from a technology and rate-based command-and-control program to an emissions trading program changes the nature of the required monitoring information. Regulating emissions under command-and-control normally only involves mandating a specific control technology and making sure it is installed and running. Under emissions trading, most facilities do not have equipment-specific control technology regulations; rather, facilities must monitor and report total emissions from all pieces of equipment and processes. In addition, the system depends on the existence of registries that can keep track of authorized emissions levels as modified by trading. Although this enhanced monitoring system provides much better information that can provide the basis for much better control over total emissions than a command-and-control system, it also costs more.

The actual enforcement records of emissions trading programs are similarly mixed, though mostly positive. For some programs like the Sulfur Allowance Program and the $NO_x$ Budget Program, coupling an effective continuous emissions monitoring program with penalties that exceed permit prices has meant complete, or near complete, compliance. As the experience in the Lead Phase-out Program demonstrates, in cases where compliance turns out to be a problem in the early years of the program, coupling an aggressive auditing program with transparency of the resulting enforcement actions can restore compliance and act as an effective deterrent.

Technology has played an important role in improving data quality and lowering the cost of the data handling involved in effective monitoring and recordkeeping.

# Lessons about Instrument Choice and Program Design

## *The Baseline Issue*

In general, tradable permit programs fit into one of two categories: credit programs or cap-and-trade programs.

Credit trading, the approach taken in the U.S. ETP (the earliest program), allows emissions reductions above and beyond baseline legal requirements to be certified as tradable credits. The baseline for credits in that program typically was provided by technology-based standards.

In a cap-and-trade program, a total emissions limit (cap) is defined and then allocated among users. Compliance is established by comparing actual use with the assigned, firm-specific cap as adjusted by any acquired or sold permits.

Establishing the baseline for credit programs in the absence of an existing permit system can be difficult. For example, the basic requirement in the Clean Development Mechanism component of the Kyoto Protocol is "additionality"—the traded reductions must be surplus to what would have been done otherwise. Deciding whether created entitlements are surplus requires establishing a baseline against which the reductions can be measured. When emissions are reduced below this baseline, the amount of the reduction that is additional can be certified as surplus.

Defining procedures that ensure that the baselines do not allow unjustified credits is no small task. A pilot program for Activities Implemented Jointly, which was established at the first Conference of the Parties in 1995 for the Kyoto Protocol, was useful for demonstrating the difficulties of ensuring additionality. Results under this program indicate that requiring a showing of additionality can impose very high transactions costs as well as introduce considerable ex ante uncertainty about the actual reductions that could be achieved.

Many credit-based programs keep a large element of the previous regulatory structure in place. For example, some programs, such as the U.S. ETP, require regulatory pre-approval for all transfers. In addition, other specific design features, such as the opt-in in the Sulfur Allowance Program, also add administrative complexity.

Theory would lead us to believe that cap-and-trade systems would be much more likely to achieve the efficiency and environmental goals than credit programs, and the evidence emerging from ex post evaluations seems to support that conclusion. This is of considerable potential importance in climate change policy since only one of the three Kyoto programs (Emissions Trading) is a cap-and-trade program.

## The Legal Nature of the Entitlement

Although the popular literature frequently refers to the tradable permit approach as "privatizing the resource," in most cases it does not actually do that. Rather, it privatizes the right to access the resource to a pre-specified degree.

Economists have argued consistently that tradable permits should be treated as secure property rights to protect the incentive to invest in the resource. Confiscation of rights or simply failing to provide adequate security for rights could undermine the entire process.

The environmental community, on the other hand, just as consistently has argued that the air belongs to the people and as a matter of ethics should not become private property. In this view, no end can justify the transformation of a community right into a private one.

The practical resolution of this conflict in most U.S. emissions trading settings has been to attempt to give "adequate" (as opposed to complete) security to the permit holders, while making it clear that permits are not property rights.[1] For example, according to the title of the Clean Air Act dealing with the Sulfur Allowance Program: "An allowance under this title is a limited authorization to emit sulfur dioxide. . . . Such allowance does not constitute a property right" (104 Stat 2591).

In practice, this means that although administrators are expected to refrain from arbitrarily confiscating rights, as sometimes happened with banked credits in the early U.S. Emissions Trading Program, they do not give up their ability to adopt a more stringent cap as the need arises. In particular, they would not be required to pay compensation for withdrawing a portion of the authorization to emit as they would if allowances were accorded full property right status. It is a somewhat uneasy compromise, but it seems to have worked.

## Adaptive Management

One of the initial fears about tradable permit systems was that they would be excessively rigid, particularly in light of the need, described immediately above, to provide adequate security to permit holders. Policy rigidity was seen as possibly preventing the system from responding either to changes in the resource base or to better information.

Existing emissions trading programs have responded to this challenge in different ways, depending on the type of resource being covered. In air pollution, the classic case involves the control of ozone-depleting gases. Soon after the establishment of a binding cap, research discovered that the problem was developing more rapidly than previously thought and that the already-established cap was inadequate. Soon after this discovery, the cap was made more stringent; the use of caps did not preclude an ability to adjust their magnitude to changing conditions.

While establishing a specific, approved process for changing the cap at the outset is one way to build adaptive management into the system, another involves a different design. Though currently not used in emissions trading, it is commonly used in biological systems, such as fisheries, where the need for cap revision may be both more regular and more frequent.

In those systems, the rights typically are defined as a share of the cap (in this case a harvesting quota). Share systems make it easier for fisheries managers to change the cap in response to changing biological conditions without triggering legal recourse by the right holder.[2] Though share rights have not been used in air pollution control, they remain a viable option.

## Caps and Safety Valves

Although caps provide a higher degree of control over aggregate emissions than traditional command-and-control, in the face of "shocks" a cap can lead to politically unacceptable cost increases. For example, RECLAIM participants experienced a very large unanticipated demand for power that could only be accommodated by increasing the output from older, more polluting plants; emissions rose considerably as a result, and permit prices soared in a way that was never anticipated.

This experience with price shocks not only provides a graphic illustration of the problem, it also demonstrates how it might be handled. The general prescription is to allow a safety valve in the form of a predefined penalty that would be imposed on all emissions over the cap. This penalty would be lower than the normal sanction imposed for noncompliance during normal situations (when compliance would be much easier). In effect, this penalty would set a maximum price that would be incurred in pursuit of environmental goals in unusually trying times.

In the case of RECLAIM, when permit prices went over a predefined threshold, the program was suspended until they figured out what to do. An alternative (substantial) fee per ton was imposed in the interim. The revenue was used to subsidize additional, alternative emissions reductions, typically from sources not covered by the cap.

## Initial Allocation Method

The initial allocation of entitlements is perhaps the most controversial aspect of any tradable permit system, including emissions trading. The four possible methods for allocating initial entitlements are: (1) random access (lotteries); (2) first come, first served; (3) administrative rules based upon eligibility criteria; and (4) auctions.

Each of these four methods have been used in one tradable permit context or another. Both lotteries and auctions frequently are used in allocating hunting permits for big game. Lotteries are more common in allocating permits among residents, and auctions are more common for allocating permits to nonresidents. First come, first served historically was common for water, especially when it was abundant.

Though an infinite number of possible administrative rules for distributing permits exist, rules that pay some, but not exclusive, attention to prior use tend to predominate. Under virtually all implemented tradable permit programs discussed in this book, existing sources get free allocations of rights rather than having to pay for them, as in an auction. Existing sources only have to purchase any additional permits they need over and above the initial allocation, as opposed to having to purchase all permits in an auction market.

Free distribution has its advantages and disadvantages. Recent work examining how the presence of pre-existing distortions in the tax system affect the efficiency of the chosen policy instrument suggest that the ability to recycle the revenue from the sale of these permits (rather than give it to users) can enhance the efficiency of the system by a large amount. That work, of course, supports the use of revenue-raising instruments such as taxes or auctioned permits rather than free distribution.

How revenues are distributed, however, also affects the relative attractiveness of alternative approaches to environmental protection from the point of view of the various stakeholders. To the extent that stakeholders can influence policy choice, using free distribution in general and prior use in particular as allocation criteria has increased the implementation feasibility of emissions trading.

This historical experience, however, need not be decisive for the future since the empirical evidence suggests that in the case of greenhouse gas control the amount of revenue needed to hold users harmless during the change is only a fraction of the total revenue available from auctioning. Allocating all permits free of charge, therefore, is not inevitable in principle, even when political feasibility considerations affect the design.

Under the typical U.S. free distribution system, new firms typically have to purchase all permits, while existing firms get an initial allocation for free. Although reserving some free permits for new firms is possible, this option rarely was exercised in practice until recently with the appearance of the national plans of Member States for complying with the EU ETS. This differential treatment of sources imposes a greater financial burden on new sources than on existing sources. In air pollution control, the evidence suggests that placing this higher, relative financial burden on new sources has retarded the introduction of new facilities and new technologies by reducing the cost advantage of building new facilities that embody the latest innovations.[3]

Basing the initial allocation on prior use also can promote inefficient strategic behavior. When sources are aware that the initial allocation will be based upon historic use, an incentive is created to inflate historic use to qualify for a larger initial allocation. This strategic behavior can intensify the degradation of the resource before the control mechanism has been established. In general, emissions trading programs typically have minimized this effect by basing initial allocations on a combination of activity levels, which are historically based, and emission rates per unit of activity, which are based on standard norms.

Compromises designed to ensure the political feasibility of the system have in some cases, particularly in RECLAIM and the Kyoto Emissions Trading Program, resulted in a cap that results in more emissions, at least initially, than would have been the case with direct regulation. This erosion of the cap can be avoided by securing agreement at the outset on the use of a ratchet mechanism. When the sum of negotiated allocations turns out to be greater than the predefined cap, a ratchet mechanism pro-rates the negotiated allocations downward until they are compatible with the predefined cap rather than accept the overage.

In the climate change case, "hot air"—the part of an Annex I country's assigned amount that is likely to be surplus to its needs without any additional efforts to reduce emissions—has been a primary concern. Hot air resulted from the initial allocation under the Kyoto Protocol because assigned amounts are defined in terms of 1990 emissions levels and for some countries, most notably Russia and Ukraine, economic contraction has resulted in substantially lower emissions levels. Hence, these countries will have surplus permits to sell, resulting in the need for less emissions reduction from new sources.

Finally, some systems allow agents other than those included in the initial allocation to participate through an opt-in procedure. This prominent feature of the Sulfur Allowance Program was plagued by adverse selection problems since the private incentive to opt-in was not compatible with efficiency. Adverse selection is not inevitable with opt-in procedures, but it is sufficiently important that it must be considered when opt-in procedures are designed.

## *Transferability Rules*

One historic concern with transferability relates to possible external effects from the transfer. Traditional theory presumes that the commodity being traded is homogeneous. Transfers of homogeneous commodities increase net benefits by allowing permits to flow to their highest valued use. In practice, however, without homogeneity transfers can confer external benefits or costs on third parties, resulting in allocations that do not maximize net benefits. The design of emissions trading systems should take these effects into consideration.

An example of an external effect involves pollutants where the location of the emission, not only the amount of emission, matters. Spatial issues arise whenever the transfer could alter the point of emission. Incorporating source location into an emissions trading program so as to deal with these spatial issues is a difficult, but manageable, proposition.

Although ambient permit markets (a special form of emissions trading) are theoretically the optimal way to address spatial issues, in practice they have not been used because of their inherent complexity. In addition, they are inconsistent with statutory prohibitions against trades that allow emissions increases, since ambient permit markets allow more emissions than other market approaches.

One way to cope with spatial issues is to use a standard emissions trading system with a more stringent cap. The difficulty with this approach is that it not only typically results in very high control costs but also provides inadequate protection against hot spots, pollutant concentrations that exceed the ambient standards at one or more points within the airshed. The high cost is a consequence of the over-control of distant sources that is necessary to ensure that the cap is stringent enough to meet a specified concentration target. The possibility of hot spots arises from the fact that this approach focuses on emissions, not concentrations.

Another possibility involves dividing the control region into zones. Zonal permit systems that can be initiated with reasonable amounts of information typically are not very effective. Since in this system permits cannot be traded across zonal boundaries, the cost penalty can be very sensitive to the initial allocation of zonal caps. Studies suggest that no conventional rule of thumb for allocating the required emissions reduction among zones comes close to the cost-effective allocation.

In practice, one common approach to resolving spatial concerns involves a system of directional trading. In directional trading, one zone is allowed to acquire permits from another zone but not vice versa. To prevent any shifting of pollution from the upwind to the downwind zone, sources in the upwind zone are prohibited from acquiring emissions permits from downwind sources, but downwind sources can acquire permits from both upwind or downwind sources. The RECLAIM system follows this approach.

An alternative strategy, also now common, involves the creation of trading rules that govern individual transactions. Trading rules are, in principle, able to reduce cost penalties below levels associated with command-and-control levels while affording greater protection from hot spots.

One trading rule strategy, the nondegradation offset, has been implemented in the United States to prevent sulfur hot spots. Known locally as regulatory tiering, this approach applies more than one regulatory regime at a time. In the Sulfur Allowance Program, sulfur emissions are controlled both by the regulations designed to achieve local ambient air quality standards as well as by the sulfur allowance trading rules. All transactions have to satisfy both programs. Thus trading is not restricted by spatial considerations (national one-for-one trades are possible), but the use of acquired allowances is subject to local regulations protecting the ambient standards. Unlike hot spot prevention programs that restrict all transactions or employ a much more strict cap, this approach prohibits only the few transactions that would result in a hot spot. Ex ante empirical analysis of this approach suggests that regulatory tiering may well be an effective compromise.

## The Temporal Dimension

Ex post evaluations have revealed that the temporal aspects of emissions trading provisions have been important in terms of both cost savings and promoting earlier reductions. Fortunately, experience with existing programs provides some specific guidance for design.

**Banking and Borrowing**. If the environmental target is appropriately defined in terms of emissions reductions (rather than ambient concentrations), firms can be allowed considerable temporal flexibility without posing an environmental risk. If, however, the goal is defined in terms of ambient pollutant concentrations, trading that allows shifts in emissions from one time period to another could lead to a clustering of emissions. Emissions that are concentrated in time lead to higher peak concentrations (hot spots) than those that are more temporally dispersed.

Emissions trading systems can incorporate temporal flexibility by allowing banking, borrowing, or both. Banking means holding a permit beyond its designated date for later use or sale. Borrowing means using a permit before its designated date.

The economic case for this flexibility is that it allows sources to optimally time their abatement investments. Flexibility in timing is important not only for reasons that are unique to each firm but also for reasons that relate to the market as a whole. When everyone makes control investments at the same time, it strains the supply capacity of the system and prices will be unnecessarily high.

When only the aggregate level of emissions matters, the price of permits would normally rise at the rate of interest and the holders would automatically choose to use them in the manner that minimizes the present value of abatement costs. Decentralized decisionmaking in this case would be compatible with social objectives. Special temporal controls would be counterproductive.

When a single, cumulative emissions cap is not sufficient to protect against damage from concentration peaks, timing must be considered as well. Situations where the damaging effects of peak concentrations are important open the door to a potentially important market failure. While firms have an incentive to minimize the present value of abatement costs, they do not have an efficient incentive to minimize the present value of all costs, including the damage caused by emissions. In general, the resulting incentive is to delay abatement, abating too little during the early periods and concentrating too much later.

Delaying abatement, however, is not always the optimal choice for the firm, even in an unrestricted permit market. When marginal abatement costs rise over time, marginal production costs fall, emissions targets decline, or output prices rise, firms have an incentive to bank, rather than borrow, permits. This was the case for the Sulfur Allowance Program. In this case, banking reduced emissions early (when concentrations were high) and increased them later (when concentrations were lower). In this case, banking clearly reduced pollution damage.

When a timing market failure is likely, it can be corrected by elements of program design. Specific examples of design modifications include: eliminating borrowing as an option, introducing increasingly more stringent aggregate emissions targets over time, or discounting borrowed (not banked) permits.

Existing tradable permit systems differ considerably in how they treat banking or the role of forward markets. No existing program is fully temporally fungible. Older pollution control programs had a more limited approach. The U.S. ETP allowed banking but not borrowing. The Lead Phase-out Program originally allowed neither, but part way through the program allowed banking. The Sulfur Allowance Program has banking but not borrowing, and RECLAIM has an overlapping timeframe for compliance that is equivalent to a highly restricted banking and borrowing system. Banking of excess reductions for future years is allowed within the first phase of the EU ETS, but banking between the 2005–2007 start-up phase and the 2008–2012 commitment period is at the discretion of member states.

**Controlling Short-Term Concentration Peaks**. Ambient standards usually are stated in terms of a long-term average, such as an annual average, or in terms of permissible exceedances of maximum, short-term average readings, such as a three-hour average, or both.

Meeting the short-term standards cost-effectively means controlling the within-year timing as well as the annual quantity of emissions. Two different types of permit systems are needed to meet two different types of situations cost-effectively.

• Periodic permits can be used to control short-term pollution peaks caused by regular, anticipated seasonal or diurnal variations in meteorological conditions.

• Episode permits can be used to control pollution during those rare, but potentially devastating, thermal inversions that can be anticipated only a day or so in advance.

Though few studies have incorporated these temporal aspects of pollution, the available evidence suggests that significant cost savings may be possible from employing both periodic and episode control permits where necessary. The more stringent the short-term standard, the larger the potential cost savings. These studies also have found that a constant control policy based on a worst-case condition frequently is not sufficient to avoid violating the ambient standards. Because the true worst case depends on emissions patterns as well as meteorological conditions, the ambient standards cannot be protected with complete assurance whenever a typical constant-control strategy is adopted.

While older emissions trading programs took no account of seasonal factors in controlling pollutants, newer programs such as the $NO_x$ Budget Trading Program represent a significant change. Under this program, the Ozone Transport Commission states established an ozone season, $NO_x$ budget cap for each affected state. Recognizing the importance of seasonality in ozone formation, this approach caps emissions only from May 1 to September 30.

While establishing seasonal controls could in principle increase emissions loadings significantly in periods other than peak periods, this emissions shifting can be countered through the use of annual average standards, complemented with permit systems tailored to those averages.

Although the level of emissions reductions in most industrialized countries has been sufficient to reduce, but not eliminate, the need for episode control, many developing countries cannot make the same case. The establishment of an episodic emissions trading system would make it easier to identify those sources that can undertake emissions reductions relatively cheaply on short notice. The current absence of this type of system creates biased incentives in the types of control adopted toward those with high fixed and low variable costs, a condition that reduces the ability of state control authorities to secure additional reductions when needed. Episodic control also could stimulate the development of new, relatively inexpensive, short-response control technologies.

## Market Power

Although market power has not been a problem with operational emissions trading programs, several design features of other aspects of the emissions trading programs could influence the likelihood of market power. These include the rule used to distribute permits, restrictions on permit transfers, and the treatment of permits freed up due to plant closures. When those design features are considered, the potential market power consequences should be considered.

Should market power seem to be a problem in a specific application, remedies such as incentive-compatible auctions, set asides, and accumulation limits are available. By requiring successful bidders to pay the price proposed by the highest rejected bidder or bidders (as opposed to paying their own bid), incentive-compatible auctions eliminate the incentive to manipulate permit prices. Set-asides provide an additional source of permits, usually controlled by the government, that limit a price-setter's power by providing buyers with an alternative. Finally, accumulation limits, which are widely use in fisheries, but not in air pollution, prevent the strategic accumulation of permits.

## Transactions Costs

The magnitude of both transactions costs and administrative costs may be affected by the emissions trading program design.

Credit-based programs, such as the U.S. ETP, typically involve a considerable amount of regulatory oversight at each step of the process (e.g., certification of credits and approval of each trade). In contrast, cap-and-trade systems rarely require either of the steps, using instead a system that compares actual and authorized emissions at the end of the year. Choosing a cap-and-trade program rather than a credit program typically can lower transactions costs.

Price transparency (making prices public), as in the Sulfur Allowance Program, can reduce the uncertainty associated with trading and facilitate negotiations about price and quantity. Furthermore, is the availability of organized exchanges where buyers and sellers can meet and knowledgeable brokers can lower the search costs for those seeking trades.

The experience of RECLAIM also demonstrates that providing mechanisms for sharing information on available technologies can reduce duplication of effort for larger firms and provide a larger menu of control options for smaller firms.

## Monitoring and Enforcement

Formal enforcement mechanisms may be based upon either criminal or civil procedures.

- Criminal sanctions usually are reserved for those situations in which a person or facility has knowingly and willfully violated the law, such as submitting a false emissions report, or has otherwise committed a violation for which society has chosen to impose the most serious legal sanctions available.

- Civil sanctions may be either administrative (i.e., directly imposed by the enforcement program) or judicial (i.e., imposed by a court or other judicial authority). When administrative enforcement mechanisms are available, civil judicial responses typically are reserved for use against more serious or recalcitrant violators, when precedents are needed, or where prompt action is important to shut down an operation or to stop an activity.

Compliance with the rules of an emissions trading system generally requires surrendering sufficient permits during a reconciliation period to cover the amount of actual emissions during that period. For this reconciliation to be reliable, an accurate official record of authorized emissions and the permit holdings that underlie them must be maintained. Recent emissions trading systems have taken advantage of the availability of new software to make data entry and data handling in these registries easier and cheaper.

In emissions trading where the number of firms is small, noncompliance on the part of even one firm in the market may affect the pollutant discharge levels of other firms through its impact on the equilibrium permit price. Because a non-complying firm needs fewer permits, noncompliance is likely to lead to a lower permit price, leading other firms to purchase more, thereby offsetting, to some extent, the effects on aggregate noncompliance.

One of the design implications of this price-induced, offsetting reaction is that a uniform, as opposed to targeted, monitoring and enforcement strategy minimizes aggregate noncompliance, given a specific monitoring and enforcement budget. Treating each firm the same regardless of its characteristics is quite different than for emissions standards, where the optimal strategy involves targeting.

Complete compliance can be achieved by ensuring that: (1) the expected cost of noncompliance (the probability of being detected times the sum of the fines for underreporting emissions and excess emissions) is higher than the permit price; and (2) the fine per unit of excess emissions is higher than the permit price. These two conditions not only provide firms with the proper incentive to submit truthful reports of their emissions, since the expected cost of lying exceeds the cost of acquiring additional permits, but it also provides the proper incentive for each firm to hold enough permits to cover their emissions. Given the incentive to submit truthful reports (from the first condition), the firms would always choose to hold enough permits if the price of acquiring those permits is less than or equal to the per-unit fine for having too few.

In practice, truthfulness in self-reporting is promoted through specifying a designated representative for each reporting source. This representative, who would assume responsibility for the submitted reports, would be subject to criminal sanctions for false reporting.

Third-party verification also can be used to improve the reliability of self-reported estimates of emissions. The EU ETS is launching a different form of monitoring activity in which externally verified self-reporting is being used in place of continuous emissions monitors, the typical form of monitoring in the United States. This form, which is analogous to existing systems of financial reporting, could prove to be more appropriate for the heterogeneous sources of greenhouse gas emissions, especially those for which CEM is not practical.

Since emissions trading usually can reduce costs considerably, using part of those gains to finance monitoring and enforcement through a tax on emissions allowances is an idea whose time may have come. Transferring the financial burden of more effective monitoring and enforcement to the sources seems to have a certain political resonance.

Penalties for excess emissions in modern emissions trading typically have two components: (1) a financial penalty on excess emissions that is larger than the permit price; and (2) the subtraction of any source's excess emissions in one commitment period from the source's allocation in the subsequent commitment period.

This review also found that small companies may not be able to offset the additional monitoring and recordkeeping costs associated with emissions trading through abatement cost reductions in the same way that larger companies can. Successfully integrating small companies into the monitoring and enforcement component of emissions trading programs requires the development of monitoring and enforcement procedures that are both effective and affordable at that scale.

## Lessons for Theory-Based Expectations

Theory creates expectations, and, in the case of tradable permits, the expectations have been high, sometimes unreasonably high. Several assumptions behind the theory may be violated in practice.

The early theory held that a cost-effective allocation ultimately would be achieved regardless of the initial allocation of permits. Whatever excess costs existed after the initial allocation would be eliminated by trading. In principle, this allows equity goals to be pursued via the initial allocation and cost-effectiveness goals to be handled by transfers.

It is now clear, however, that in several circumstances the degree of cost-effectiveness achieved by emissions trading is dependent upon the initial allocation, not independent of it. Specific examples of this dependence are illustrated by: (1) the double dividend issue, where recycled revenue from auctions could reduce the welfare costs; (2) the differential treatment of new sources under grandfathering; (3) creating market power by allocating too many permits to a particular firm or set of firms; (4) strategic considerations, where firms increase their emissions to qualify for more permits; and (5) transactions costs, where trading is sufficiently inhibited that the ability of the market to overcome any cost-effectiveness deficiencies in the initial allocation is precluded.

Theoretical models also may fail to incorporate sufficient complexity when they assume that the tradable commodity is homogeneous. In many practical applications, the tradable commodity clearly is not homogenous. The location or timing of permit use may matter, as might the specific control methods used by the permit holder. The impact of nonhomogeneity is intensified when the associated environmental benefits or damages are external to the users. In this case, permit holders who use or trade permits cannot be expected to minimize society's costs when they minimize their own.

Another aspect of tradable permit systems that seems to have been underappreciated is endogeneity. The choice of a policy instrument can affect aspects of implementation that frequently are considered exogenous to the analysis but in fact are key aspects of choice. These include the targeted degree of control, the feasibility of implementation, the likelihood of compliance, and the form and intensity of monitoring and enforcement, as well as the degree of technical change.

The role that fairness or ethical considerations play in the design of operating tradable permit systems seems to be more important than typically believed. Analysis that assumes that fairness is either completely handled by the initial allocation or has no analytical importance may miss a comparative aspect of policy instrument choice that seems to matter in practice.

Permit markets certainly have achieved a large and growing niche in the collection of favored policies to control air pollution. This reading of the evidence suggests that is appropriate, but it also suggests that the resource context and program design not only matter, they matter a lot.

## Lessons About Ex Post Evaluation

This review suggests that the form of the ex post evaluation matters in terms of what can be learned from it. Efficiency studies, for example, that consider the programmatic effects on other markets, particularly in the presence of distortionary taxes, find a marked advantage for auctions, due to the ability to recycle revenue, over initial allocations based on free distribution. Cost-effectiveness studies, which are unable to consider this aspect, find no such advantage.

Emissions trading, of course, usually is not implemented in a vacuum and that complicates the process of sorting out the specific consequences of implementing the program. Emissions trading frequently complements other policies.

- The U.S. Sulfur Allowance Program operates within the more general framework of sulfur oxide regulation established by the National Ambient Air Quality Standards.

- The RECLAIM program in California has been affected by dramatic developments in that state's electric deregulation program.

- The Chicago VOM program operated within a large array of traditional regulations that dictated many of the choices made by sources.

- The Santiago, Chile, program sought to reduce emissions at the exact time that natural gas became available for the first time.

The interdependence of these programs makes it difficult to disentangle the unique effects of emissions trading and to draw implications for how the policy might work in a different policy environment.

Failure to recognize either nonhomogeneity or endogeneity in the evaluation process can lead to biased evaluations. Considering aspects such as the feasibility of the system, the level of the target, the likelihood of reaching the target, and the effectiveness of monitoring and enforcement as outside the scope of analysis can miss important consequences of instrument choice.

Historically, tradable permit systems have tended to evolve considerably over time. Regulators have had to become comfortable with the flexibility these systems afford. Users have had to become comfortable with the fact that defining the means of control is now up to them. Initial tradable permit markets may bear only a remote resemblance to more familiar goods or asset markets.

Failure to recognize the evolutionary nature of the system may result in conclusions drawn from an analysis of a transitory stage being mistakenly interpreted as reflecting what would have been found at a later stage. Early evaluations may not provide much insight about the ultimate success or failure of the program, since dramatic change is so common.

Failure to think seriously about the timing of the evaluation also can lead to an underestimation of the importance of early evaluations in shaping the speed and form of the evolution. Since evolution is so common for emissions trading, a strong case can be made for thinking of mid-course corrections as routine.

## Concluding Comments

Emissions trading seems to provide a good example of the pendulum theory of public policy. In the early 1970s, emissions trading was considered an academically intriguing, but ultimately impractical, idea. It had trouble getting on the national agenda. Reformers had few successes.

However, that changed once the Lead Phase-out and Sulfur Allowance Programs became law and demonstrated not only the feasibility of the approach but also its power. Emboldened by success, expectations and enthusiasm started to outrun reality.

In the final stage, the one I believe we are in now, reality once again is reasserting itself. Both policymakers and academics are beginning to realize not only that emissions trading has achieved a considerable measure of success but also that it has specific, identifiable weaknesses.

We also have learned that not all emissions trading programs are equal. Some designs are better than others. Furthermore, one size does not fit all. Emissions trading programs can and should be tailored to each specific application.

The evidence suggests that while emissions trading is no panacea, well-designed programs that are targeted at pollution problems appropriate for this form of control are beginning to occupy an important and durable niche in the evolving menu of environmental policies. The remaining task is to continue refining our understanding of the parameters of that niche.

# Notes

1. One prominent exception to the general rule that tradable permits are not full property rights can be found in the New Zealand Individual Transferable Quotas fisheries management system. It grants full property rights in perpetuity (National Research Council Committee to Review Individual Fishing Quotas 1999, 97).

2. Some fisheries and water allocation systems actually create two different, but related rights. The first conveys the share of the cap, while the second conveys the right to withdraw a specified amount in a particular year. Obviously, the second right is derived from the first, but separating the two rights allows a user to sell the current access right (perhaps due to an illness or malfunctioning equipment) without giving up the right of future access embodied in the share right.

3. The differential treatment of new sources is, of course, not unique to emissions trading. Conventional regulation also imposes more stringent requirements on new sources than existing ones.

# References

Åhman, M., D. Burtraw, J. Kruger, and L. Zetterberg. 2005. The Ten-Year Rule: Allocation of Emissions Allowances in the EU Emissions Trading System. Discussion paper 05-30. Washington, DC: Resources for the Future.

Anderson, Robert J., Jr., Robert O. Reid, and Eugene P. Seskin. 1979. An Analysis of Alternative Policies for Attaining and Maintaining a Short-Term $NO_2$ Standard, a report to the Council on Environmental Quality. Princeton, NJ: MATHTECH.

Anderson, Robert C. 2001. The United States Experience with Economic Incentives for Pollution Control. Report EPA-240-R-01-001. Washington, DC: National Center for Environmental Economics.

Atkinson, Scott E. 1983. Marketable Pollution Systems and Acid Rain Externalities. *Canadian Journal of Economics* 16(4): 704–722.

———. 1994. Tradable Discharge Permits: Restrictions on Least Cost Solutions. In *Economic Instruments for Air Pollution Control,* edited by G. Klaassen and F. R. Førsund. Boston, MA: Kluwer Academic Publishers, 3–21.

Atkinson, S.E., and Donald H. Lewis. 1974. A Cost-Effectiveness Analysis of Alternative Air Quality Control Strategies. *Journal of Environmental Economics and Management* 1(3): 237–250.

Atkinson, S.E., and B.B.J. Morton. 2004. Determining the Cost-effective Size of an Emission Trading Region for Achieving an Ambient Standard. *Resource and Energy Economics* 26(3): 295–315.

Atkinson, S.E., and T.H. Tietenberg. 1982. The Empirical Properties of Two Classes of Designs for Transferable Discharge Permit Markets. *Journal of Environmental Economics and Management* 9(2): 101.

———. 1984. Approaches for Reaching Ambient Standards in Non-Attainment Areas: Financial Burden and Efficiency Considerations. *Land Economics* 60(2): 148–159.

———. 1987. Economic Implications of Emission Trading Rules for Local and Regional Pollutants. *Canadian Journal of Economics* 20(2): 370–386.

———. 1991. Market Failure in Incentive-Based Regulation: The Case of Emissions Trading. *Journal of Environmental Economics and Management* 21(1): 17–31.

Babiker, M. 2005. Climate Change Policy, Market Structure, and Carbon Leakage. *Journal of International Economics* 65(2): 421–445.

Baumol, W.J., and W.E. Oates. 1971. The Use of Standards and Prices for Protection of the Environment. *Swedish Journal of Economics* 73: 42–54.

Becker, Gary S. 1968. Crime and Punishment: An Economic Analysis. *Journal of Political Economy* 76(2): 169–217.

Beder, S. 1996. Charging the Earth—the Promotion of Price-Based Measures for Pollution-Control. *Ecological Economics* 16(1): 51.

Ben-David, S. et al. 2000. Attitudes Toward Risk and Compliance in Emission Permit Markets. *Land Economics* 76(4): 590–600.

Berland, H., D.J. Clark, and P.A. Pederson. 2001. Rent Seeking and the Regulation of a Natural Resource. *Marine Resource Economics* 16: 219-233.

Berman, E., and L.T.M. Bui. 2001. Environmental Regulation and Labor Demand: Evidence from the South Coast Air Basin. *Journal of Public Economics* 79(2): 265–295.

Bernard, A. et al. 2003. Russia's Role in the Kyoto Protocol. Report No. 98. Cambridge, MA: MIT Joint Program on the Science and Policy of Global Change.

Bohi, D.R., and D. Burtraw. 1997. Trading Expectations and Experience. *The Electricity Journal* 10(7).

Bohm, P., and B. Carlén. 1999. Emission Quota Trade Among the Few: Laboratory Evidence of Joint Implementation Among Committed Countries. *Resource and Energy Economics* 21: 43–66.

Bovenberg, A.L. and L.H. Goulder. 2000. Neutralizing Adverse Impacts of $CO_2$ Abatement Policies: What Does it Cost? In *Behavioral and Distributional Effects of Environmental Policy*, edited by C.E. Carraro and G.E. Metcalf. Chicago: University of Chicago Press.

Boyd, J.D. 1993. Mobile Source Emissions Reduction Credits as a Cost-Effective Measure for Controlling Urban Air Pollution. In *Cost-Effective Control of Urban Smog*, edited by R.F. Kosobud, W.A. Testa, and D.A. Hanson. Chicago: Federal Reserve Bank of Chicago, 149.

Brady, Gordon L., and Richard E. Morrison. 1982. Emission Trading: An Overview of the EPA Policy Statement. Policy and Research Analysis Report 82-2. Washington, DC: National Science Foundation.

Bruneau, J.F. 2004. A Note on Permits, Standards, and Technological Innovation. *Journal of Environmental Economics and Management* 48(3): 1192–1199.

Burniaux, J.M. 1999. How Important is Market Power in Achieving Kyoto? An Assessment Based on the GREEN Model. Paris: Organisation for Economic Co-operation and Development.

Burtraw, D. 1996. The $SO_2$ Emissions Trading Program: Cost Savings without Allowance Trades. *Contemporary Economic Policy* XIV(2): 79.

Burtraw, D. and E. Mansur. 1999. Environmental Effects of $SO_2$ Trading and Banking. *Environmental Science and Technology* 33(20): 3489–3494.

Burtraw, D. et al. 1998. Improving Efficiency in Bilateral Emission Trading. *Environmental and Resource Economics* 11(1): 19–33.

Burtraw, D., and K. Palmer. 2004. $SO_2$ Cap-and-Trade Program in the United States: A "Living Legend" of Market Effectiveness. In *Choosing Environmental Policy: Comparing Instruments and Outcomes in the United States and Europe*, edited by W.

Harrington, R.D. Morgenstern, and T. Sterner. Washington, DC: Resources for the Future, 41–66.

Carlén, B. 2003. Market Power in International Carbon Emissions Trading: A Laboratory Test. *Energy Journal* 24(3): 1–26.

Carlson, Curtis, Dallas R. Burtraw, Maureen Cropper, and Karen L. Palmer. 2000. Sulfur Dioxide Control by Electric Utilities: What Are the Gains From Trade? *Journal of Political Economy* 108(6): 1292–1326.

Carraro, C.E. 2002. Climate Change Policy: Models, Controversies and Strategies. In *The International Yearbook of Environmental and Resource Economics 2002/2003*, edited by T. Tietenberg and H. Folmer. Cheltenham, UK: Edward Elgar Publishing, 1-65.

Cason, T.N. 1993. Seller Incentive Properties of EPA's Emission Trading Auction. *Journal of Environmental Economics and Management* 25(2): 177.

Cason, T.N., and L. Gangadharan. 1998. An Experimental Study of Electronic Bulletin Board Trading for Emission Permits. *Journal of Regulatory Economics* 14(1): 55–73.

———. 2005. *Emissions Variability in Tradable Permit Markets with Imperfect Enforcement and Banking.* West Lafayette, IN: Purdue University.

Cason, T.N. et al. 2003. A Laboratory Study of Auctions for Reducing Non-point Source Pollution. *Journal of Environmental Economics and Management* 46(3): 446–471.

Chavez, C.A., and J.K. Stranlund. 2003. Enforcing Transferable Permit Systems in the Presence of Market Power. *Environmental and Resource Economics* 25(1): 65–78.

Chayes, A., and A.H. Chayes. 1993. On Compliance. *International Organization* 47: 175-205.

Coase, R. 1960. The Problem of Social Cost. *Journal of Law and Economics* 3 (October): 1-44.

Coggins, J.S., and V.H. Smith. 1993. Some Welfare Effects of Emission Allowance Trading in a Twice-Regulated Industry. *Journal of Environmental Economics and Management* 25(3): 275-297.

Collinge, R.A., and W.E. Oates. 1982. Efficiency in Pollution-Control in the Short and Long Runs: A System of Rental Emission Permits. *Canadian Journal of Economics* 15(2): 346–354.

Conrad, K., and R.E. Kohn. 1996. The U.S. Market for $SO_2$ Permits—Policy Implications of the Low Price and Trading Volume. *Energy Policy* 24(12): 1051.

Copeland, B.R., and M.S. Taylor. 2004. Trade, Growth, and the Environment. *Journal of Economic Literature* XLI: 7-71.

Council on Environmental Quality. 1980. Environmental Quality: The Eleventh Annual Report. Washington, DC: U.S. Government Printing Office.

Crandall, Robert W. 1983. *Controlling Industrial Pollution: The Economics and Politics of Clean Air.* Washington, DC: Brookings Institution.

Crandall, Robert W., and Paul R. Portney. 1984. The Environmental Protection Agency in the Reagan Administration. In *Natural Resources and the Environment*, edited by Paul R. Portney. Washington, DC: Urban Institute Press, 47–81.

Crocker, T.D. 1966. The Structuring of Atmospheric Pollution Control Systems. In *The Economics of Air Pollution*, edited by H. Wolozin. New York: W.W. Norton & Co, 61–86.

Cronshaw, M., and J.B. Kruse. 1996. Regulated Firms in Pollution Permit Markets with Banking. *Journal of Regulatory Economics* 9: 179–189.

Dales, J.H. 1968. *Pollution, Property and Prices.* Toronto: University of Toronto Press.

David, M., W. Eheart, E. Joeres, and E. David. 1980. Marketable Permits for the Control of Phosphorus Effluent into Lake Michigan. *Water Resources Research* 16(2): 263–270.

Dean, J.M. 1992. Trade and the Environment: A Survey of the Literature. In *International Trade and the Environment,* edited by P. Low. Washington, DC: The World Bank, 15-28.

de Lucia, Russell J. 1974. An Evaluation of Marketable Effluent Permit Systems, Report no. EPA-60015-74-030. Washington, DC: U.S. Environmental Protection Agency.

Delache, Xavier. 2003. Ex Post Evaluation in France: Framework and Lessons Learned. Proceedings of the OECD Workshop on Ex Post Evaluation of Tradable Permits: Methodological and Policy Issues. January 21–22, Paris.

denElzen, M.G.J., and A.P.G. deMoor. 2002. Analyzing the Kyoto Protocol under the Marrakesh Accords: Economic Efficiency and Environmental Effectiveness. *Ecological Economics* 43(2-3): 141-158.

Dinan, Terry, and D.L. Rogers. 2002. Distributional Effects of Carbon Allowance Trading: How Government Decisions Determine Winners and Losers. *National Tax Journal* 55: 199–222.

Dorfman, Nancy S., and Arthur Snow. 1975. Who Will Pay for Pollution Control? The Distribution by Income of the Burden of the National Environmental Protection Program, 1972-80. *National Tax Journal* 28(1): 101–115.

Downing, Paul, and William D. Watson. 1975. Cost-Effective Enforcement of Environmental Standards. *Journal of the Air Pollution Control Association* 25(7): 705–710.

Drayton, William. 1980. Economic Law Enforcement. *Harvard Environmental Law Review* 4(1): 1-40.

Dudek, D.J., and J. Palmisano. 1988. Emissions Trading: Why is this Thoroughbred Hobbled? *Columbia Journal of Environmental Law* 13(2): 217–256.

Dudek, D. et al. 1992. Environmental Policy for Eastern Europe: Technology-Based versus Market-Based Approaches. *Columbia Journal of Environmental Law* 17(1): 1.

Eheart, J. Wayland, E. Downey Brill, Jr., and Randolph M. Lyon. 1983. Transferable Discharge Permits for Control of BOD: An Overview. In *Buying a Better Environment: Cost-Effective Regulation Through Permit Trading,* edited by Erhard F. Joeres and Martin H. David. Madison, WI: University of Wisconsin Press, 163–195.

Ekins, P. 1996. The Secondary Benefits of $CO_2$ Abatement: How Much Emission Reduction Do They Justify? *Ecological Economics* 16(1): 13–24.

Ellerman, A. Denny. 2003. The U.S. $SO_2$ Cap-and-Trade Program. Proceedings of the OECD Workshop on Ex Post Evaluation of Tradable Permits: Policy Evaluation, Design and Reform. Paris: Organisation for Economic Co-operation and Development.

Ellerman, A.D., and J.P. Montero. 2005. The Efficiency and Robustness of Allowance Banking in the U.S. Acid Rain Program. CEEPR Working Paper 2005-005. Cambridge, MA: MIT Center for Energy and Environmental Policy Research.

Ellerman, A.D., et al. 1997. Emissions Trading Under the U.S. Acid Rain Program: Evaluation of Compliance Costs and Allowance Market Performance. Cambridge, MA: MIT Center for Energy and Environmental Policy Research.

——. 2000. *Markets for Clean Air: The U.S. Acid Rain Program.* Cambridge, UK: Cambridge University Press.

Elman, Barry. 1983. *Status Report on Emission Trading Activity.* Washington, DC: U.S. Environmental Protection Agency.

Farrell, A., et al. 1999. The $NO_x$ Budget: Market-Based Control of Tropospheric Ozone in the Northeastern United States. *Resource and Energy Economics* 21(2): 103–124.

Feldman, S.L., and R.K. Raufer. 1987. *Emissions Trading and Acid Rain: Implementing a Market Approach to Pollution Control.* Totowa, NJ: Rowman & Littlefield.

Fischer, C., and A. Fox. 2004. Output-Based Allocations of Emissions Permits: Efficiency and Distributional Effects in a General Equilibrium Setting with Taxes and Trade. Discussion paper 04–37. Washington, DC: Resources for the Future.

Fischer, C., I. Parry, and W. Pizer. 2003. Instrument Choice for Environmental Protection When Technological Innovation is Endogenous. *Journal of Environmental Economics and Management* 45: 523–545.

Førsund, F.R., and E. Nævdal. 1994. Trading Sulfur Emissions in Europe. In *Economic Instruments for Air Pollution Control,* edited by G. Klaassen and F.R. Førsund. Norwell, MA: Kluwer Academic Publishers, 231–248.

——. 1998. Efficiency Gains under Exchange-Rate Emission Trading. *Environmental and Resource Economics* 12(4): 403–423.

Foster, V., and R.W. Hahn. 1995. Designing More Efficient Markets: Lessons from Los Angeles Smog Control. *Journal of Law and Economics* 38(1): 19.

Fowlie, M., and J. Perloff. 2004. The Effect of Pollution Permit Allocations on Firm-Level Emissions. CUDARE working paper #968. Berkeley, CA: University of California.

Franciosi, R. et al. 1993. An Experimental Investigation of the Hahn-Noll Revenue Neutral Auction for Emissions Licenses. *Journal of Environmental Economics and Management* 24(1): 1–24.

Fromm, O., and B. Hansjurgens. 1996. Emission Trading in Theory and Practice: An Analysis of RECLAIM in Southern California. *Environment and Planning C: Government and Policy* 14(3): 367–384.

Fullerton, D., and G.E. Metcalf. 2002. Cap and Trade Policies in the Presence of Monopoly and Distortionary Taxation. *Resource and Energy Economics* 24(4): 327–347.

Fullerton, D. et al. 1997. Sulfur Dioxide Compliance of a Regulated Utility. *Journal of Environmental Economics and Management* 34(1): 32–53.

Gangadharan, L. 1997. *Transactions Costs in Tradable Emissions Markets: An Empirical Study of the Regional Clean Air Incentives Market in Los Angeles.* Los Angeles, CA: University of Southern California.

Garvie, D., and A. Keeler. 1994. Incomplete Enforcement with Endogenous Regulatory Choice. *Journal of Public Economics* 55: 141–162.

Gianessi, Leonard P., and Henry M. Peskin. 1980. The Distribution of the Costs of Federal Water Pollution Control Policy. *Land Economics* 56(1): 85–102.

Gianessi, Leonard P., Henry M. Peskin, and Edward Wolff. 1979. The Distributional Effects of Uniform Air Pollution Policy in the United States. *Quarterly Journal of Economics* 93(2): 281–301.

Godby, R. 2000. Market Power and Emissions Trading: Theory and Laboratory Results. *Pacific Economics Review* 5(3): 349–363.

———. 2002. Market Power in Laboratory Emission Permit Markets. *Environmental and Resource Economics* 23(3): 279–318.

Godby, R., S. Mestelman, R.A. Muller, and J.D. Welland. 1997. Emissions Trading with Shares and Coupons when Control Over Discharges is Uncertain. *Journal of Environmental Economics and Management* 32(3): 359–381.

Godby, R. et al. 1999. Experimental Tests of Market Power. In *Emission Trading Markets: Environmental Regulation and Market Power*, edited by E. Petrakis, E.S. Sartzetakis, and A. Xepapadeas. Cheltenham, UK: Edward Elgar Publishing, 67–94.

Goldenberg, E. 1993. The Design of an Emissions Permit Market for RECLAIM: A Holistic Approach. *UCLA Journal of Environmental Law & Policy* 11(2): 297.

Goodin, R.E. 1994. Selling Environmental Indulgences. *Kyklos* 47(4): 573.

Goodstein, E. 1996. Jobs and the Environment—an Overview. *Environmental Management* 20(3): 313–321.

Goulder, Lawrence, Ian Parry, and Dallas Burtraw. 1997. Revenue-Raising Versus Other Approaches to Environmental Protection: The Critical Significance of Preexisting Tax Distortions. *RAND Journal of Economics* 28: 708–731.

Goulder, Lawrence, Ian Parry, Roberton Williams III, and Dallas Burtraw. 1999. The Cost-Effectiveness of Alternative Instruments for Environmental Protection in a Second-Best Setting. *Journal of Public Economics* 72: 329–360.

Government Accounting Office. 1984. An Analysis of Issues Concerning "Acid Rain." Washington, DC: U.S. Government Accounting Office.

Hagem, C. 2003. The Merits of Non-tradable Quotas as a Domestic Policy Instrument to Prevent Firm Closure. *Resource and Energy Economics* 25(4): 373–386.

Hagem, C., and H. Westkog. 1998. The Design of a Dynamic Tradable Quota System under Market Imperfections. *Journal of Environmental Economics and Management* 36(1): 89–107.

Hahn, Robert W. 1982. Market Power and Transferable Property Rights. In *Implementing Tradable Permits for Sulfur Oxides Emissions: A Case Study in the South Coast Air Basin* (vol. 3), edited by Glen R. Cass, Robert W. Hahn, Roger G. Noll, William P. Rogerson, George Fox, and Asha Paragjape. Pasadena, CA: California Institute of Technology, C45–C70.

———. 1983. Designing Markets in Transferable Property Rights: A Practitioner's Guide. In *Buying a Better Environment. Cost-Effective Regulation Through Permit Trading*, edited by Erhard F. Joeres and Martin H. David. Madison, WI: University of Wisconsin Press, 83–97.

———. 1984. Market Power and Transferable Property Rights. *Quarterly Journal of Economics* 99(4): 753–765.

———. 1986. Trade-offs in Designing Markets With Multiple Objectives. *Journal of Environmental Economics and Management* 13(1): 1–12.

———. 1989. Economic Prescriptions for Environmental Problems: How the Patient Followed the Doctor's Orders. *The Journal of Economic Perspectives* 3(2): 95.

Hahn, R.W., and R.L. Axtell. 1995. Reevaluating the Relationship Between Transferable Property Rights and Command-and-Control Regulation. *Journal of Regulatory Economics* 8(2): 125–148.

Hahn, R.W., and G.L. Hester. 1989a. Marketable Permits: Lessons from Theory and Practice. *Ecology Law Quarterly* 16: 361–406.

———. 1989b. Where Did All the Markets Go? An Analysis of EPA's Emission Trading Program. *Yale Journal of Regulation* 6(1): 109.

Hahn, R.W., and A.M. McGartland. 1989. The Political Economy of Instrument Choice: An Examination of the U.S. Role in Implementing the Montreal Protocol. *Northwestern University Law Review* 83(3): 592–611.

Hahn, Robert W., and Roger G. Noll. 1982. Designing an Efficient Permits Market. In *Implementing Tradeable Permits for Sulfur Oxide Emissions: A Case Study in the South Coast Air Basin* (vol. II), edited by Glen R. Cass et al. Pasadena, CA: California Institute of Technology, 102–134.

Hall, J.V., and A.L. Walton. 1996. A Case Study in Pollution Markets: Dismal Science vs. Dismal Reality. *Contemporary Economic Policy* XIV(2): 67.

Harford, Jon D. 1978. Firm Behavior Under Imperfectly Enforceable Pollution Standards and Taxes. *Journal of Environmental Economics and Management* 5(1): 26–43.

Harrington, Winston. 1981. *The Regulatory Approach to Air Quality Management.* Washington, DC: Resources for the Future.

Harrington, W., R.D. Morgenstern, and T. Sterner (eds.). 2004. *Choosing Environmental Policy: Comparing Instruments and Outcomes in the United States and Europe.* Washington, DC: Resources for the Future.

Harrison, David, Jr. 1983. Case Study 1: The Regulation of Aircraft Noise. In *Incentives for Environmental Protection,* edited by T.C. Schelling. Cambridge, MA: MIT Press, 41.

———. 1994. *The Distributional Effects of Economic Instruments for Environmental Protection.* Paris: Organisation for Economic Co-operation and Development.

———. 2002. Tradable Permits for Air Quality and Climate Change. In *The International Yearbook of Environmental and Resource Economics: 2002/2003,* edited by T. Tietenberg and H. Folmer. Cheltenham, UK: Edward Elgar Publishing, 311–372.

———. 2004. Ex Post Evaluation of the RECLAIM Emissions Trading Programmes for the Los Angeles Air Basin. In *Tradable Permits: Policy Evaluation, Design and Reform.* Paris: Organisation for Economic Co-operation and Development, 45–69.

Harrison, David, Jr., and Paul R. Portney. 1982. Who Loses from Reform of Environmental Regulation. In *Reform of Environmental Regulation,* edited by Wesley A. Magat. Cambridge, MA: Ballinger, 147–179.

Hartridge, O. 2003. The UK Emissions Trading Scheme: Progress Report. Intervention at the OECD Workshop on the Ex Post Evaluation of Tradable Permits, January 21–22, Paris.

Hatcher, A. et al. 2000. Normative and Social Influences Affecting Compliance with Fishery Regulations. *Land Economics* 76(3): 448–461.

Hausker, K. 1992. The Politics and Economics of Auction Design in the Market for Sulfur Dioxide Pollution. *Journal of Policy Analysis and Management* 11(4): 553–572.

Helm, C. 2003. International Emissions Trading with Endogenous Allowance Choices. *Journal of Public Economics* 87(12): 2737–2747.

Henderson, J.V. 1996. Effects of Air-Quality Regulation. *American Economic Review* 86(4): 789–813.

Hockenstein, J.B. et al. 1997. Crafting the Next Generation of Market-Based Environmental Tools. *Environment* 39(4): 12.

Howe, C. W., and D. R. Lee. 1983. Priority Pollution Rights: Adapting Pollution Control to a Variable Environment. *Land Economics* 59(2): 141–149.

———. 1983. Organizing the Receptor Side of Pollution Rights Markets. *Australian Economic Papers* 22(41): 280–289.

Innes, R. et al. 1991. Emission Permits Under Monopoly. *Natural Resource Modeling* 5(3): 321–343.

Jacobs, James J., and George L. Casler. 1979. Internalizing Externalities of Phosphorous Discharges from Crop Production to Surface Water: Effluent Taxes versus Uniform Reductions. *American Journal of Agricultural Economics* 61(2): 309–312.

Jaffe, A.B., and R.N. Stavins. 1995. Dynamic Incentives of Environmental Regulations: The Effects of Alternative Policy Instruments on Technology Diffusion. *Journal of Environmental Economics and Management* 29(3 Suppl., Part 2): S43–S63.

Jenkins, G.P., and R. Lamech. 1992. International Market-Based Instruments for Pollution Control. *Bulletin for International Fiscal Documentation* 46(11): 523.

Johnson, Edwin L. 1967. A Study in the Economics of Water Quality Management. *Water Resources Research* 3(1): 291–305.

Johnson, S.L., and D.M. Pekelney. 1996. Economic Assessment of the Regional Clean Air Incentives Market: A New Emissions Trading Program for Los Angeles. *Land Economics* 72(3): 277.

Johnson, Warren B. 1983. Interregional Exchanges of Air Pollution: Model Types and Applications. *Journal of the Air Pollution Control Association* 33(6): 563–574.

Joskow, P.L. et al. 1998. The Market for Sulfur Dioxide Emissions. *American Economic Review* 88(4): 669–685.

Joskow, P.L., and E. Kahn. 2002. A Quantitative Analysis of Pricing Behavior in California's Wholesale Electricity Market During Summer 2000. *Energy Journal* 23(4): 1–35.

Jung, C. H. et al. 1996. Incentives for Advanced Pollution Abatement Technology at the Industry Level: An Evaluation of Policy Alternatives. *Journal of Environmental Economics and Management* 30(1): 95–111.

Kampas, A., and B. White. 2003. Selecting Permit Allocation Rules for Agricultural Pollution Control: A Bargaining Solution. *Ecological Economics* 47(2-3): 135–147.

Keeler, A. G. 1991. Noncompliant Firms in Transferable Discharge Permit Markets: Some Extensions. *Journal of Environmental Economics and Management* 21(2): 180–189.

Kelman, Steven. 1981. *What Price Incentives? Economists and the Environment.* Westport, CT: Greenwood Publishing Group.

Kennedy, P.W., and B. Laplante. 1999. Environmental Policy and Time Consistency: Emission Taxes and Emissions Trading. In *Emission Trading Markets: Environmental Regulation and Market Power*, edited by E. Petrakis, E.S. Sartzetakis, and A. Xepapadeas. Cheltenham, UK: Edward Elgar Publishing, 116–144.

Kerr, Suzi, and David Maré. 1999. Transaction Costs and Tradable Permit Markets: The United States Lead Phasedown. Unpublished manuscript available at: http://www.motu.org.nz/abstracts/transaction_costs.htm (accessed October 31, 2005).

Kerr, S., and R.G. Newell. 2003. Policy-Induced Technology Adoption: Evidence from the U.S. Lead Phasedown. *Journal of Industrial Economics* 51: 317–343.

Kete, N. 1992. The U.S. Acid Rain Control Allowance Trading System. In *Climate Change: Designing a Tradeable Permit System*, edited by T. Jones and J. Corfee-Morlot. Paris: Organisation for Economic Co-operation and Development, 78–108.

Klaassen, G. 1996. *Acid Rain and Environmental Degradation: The Economics of Emission Trading.* Cheltenham, UK: Edward Elgar Publishing.

Klaassen, G., and F.R. Førsund (eds.). 1994 *Economic Instruments for Air Pollution Control.* Boston: Kluwer Academic Publishers.

Klaassen, G., and A. Nientjes. 1997. Sulfur Trading Under the 1990 CAAA in the U.S.: An Assessment of First Experiences. *Journal of Institutional and Theoretical Economics* 153(2): 384.

Kline, J., and F. Menezes. 1999. A Simple Analysis of the U.S. Emission Permit Auctions. *Economics Letters* 65(2): 183–189.

Kling, C.L. 1994a. Emission Trading vs. Rigid Regulations in the Control of Vehicle Emissions. *Land Economics* 70(2): 174.

———. 1994b. Environmental Benefits from Marketable Discharge Permits or an Ecological vs. Economical Perspective on Marketable Permits. *Ecological Economics* 11(1): 57.

Kling, C.L., and J. Rubin. 1997. Bankable Permits for the Control of Environmental Pollution. *Journal of Public Economics* 64(1): 99–113.

Kling, C.L., and J.H. Zhao. 2000. On the Long-Run Efficiency of Auctioned vs. Free Permits. *Economics Letters* 69(2): 235–238.

Kneese, Allen V., and F. Lee Brown. 1981. *The Southwest Under Stress: National Resource Development Issues in a Regional Setting.* Baltimore, MD: Johns Hopkins University Press for Resources for the Future.

Koch, James C., and Robert E. Leone. 1979. The Clean Water Act: Unexpected Impacts on Industry. *Harvard Environmental Law Review* 3(May): 84–111.

Kosobud, R.F., H.H. Stokes, C.D. Tallarico, and B.L. Scott. 2004. The Chicago VOC Trading System: The Consequences of Market Design for Performance. Working Paper #2004-019. Cambridge, MA: MIT/CEEPR.

Kostow, Lloyd P., and John F. Kowalcyzk. 1983. A Practical Emission Trading Program. *Journal of the Air Pollution Control Association* 33(10): 982–984.

Kruger, J.A., and W.A. Pizer. 2004. Greenhouse Gas Trading in Europe—The New Grand Policy Experiment. *Environment* 46(8): 8.

Kruger, J.A. et al. 1999. A Tale of Two Revolutions: Administration of the $SO_2$ Trading Program. Draft report. Washington, DC: U.S. Environmental Protection Agency.

Krumm, R., and D. Wellisch. 1995. On the Efficiency of Environmental Instruments in a Spatial Economy. *Environmental and Resource Economics* 6(1): 87–98.

Krupnick, Alan J. 1983. Costs of Alternative Policies for the Control of $NO_2$ in the Baltimore Region. Unpublished Resources for the Future working paper.

Krupnick, A.J. et al. 1983. On Marketable Air Pollution Permits: The Case for a System of Pollution Offsets. *Journal of Environmental Economics and Management* 10(3): 233–247.

Krupnick, Alan J., Wallace E. Oates, and Eric Van de Verg. 1983. On Marketable Air-Pollution Permits: The Case for a System of Pollution Offsets. *Journal of Environmental Economics and Management* 10(3): 233–247.

Kruse, Jamie Brown, and Mark Cronshaw. 1999a. An Experimental Analysis of Emission Permits with Banking and the Clean Air Act Amendments of 1990. In *Research in Experimental Economics* (vol. 7), edited by R. Mark Isaac and Charles Holt. Stamford, CT: JAI Press, 1–24.

———. 1999b. Temporal Properties of a Market for Emission Permits. In *Research in Experimental Economics* (vol. 7), edited by R. Mark Isaac and Charles Holt. Stamford, CT: JAI Press, 181–203.

Kuhn, H.W., and A.W. Tucker. 1951. Nonlinear Programming. In *Proceedings of the Second Berkeley Symposium on Mathematical Statistics and Probability*, edited by J. Neyman. Berkeley, CA: University of California Press, 481–492.

Kuik, O., and M. Mulder. 2004. Emissions Trading and Competitiveness: Pros and Cons of Relative and Absolute Schemes. *Energy Policy* 32(6): 737–745.

Laffont, J.J., and J. Tirole. 1996a. Pollution Permits and Compliance Strategies. *Journal of Public Economics* 62(1-2): 85–125.

———. 1996b. Pollution Permits and Environmental Innovation. *Journal of Public Economics* 62(1-2): 127–140.

Larsen, Ralph I. 1971. A Mathematical Model for Relating Air Quality Measurements to Air Quality Standards. Environmental Protection Agency Office of Air Programs AP-89. Washington, DC: U.S. Government Printing Office.

Leiby, P., and J. Rubin. 2001. Intertemporal Permit Trading for the Control of Greenhouse Gas Emissions. *Environmental and Resource Economics* 19(3): 229–256.

Levinson, Michael. 1980. Deterring Air Polluters through Economically Efficient Sanctions: A Proposal for Amending the Clean Air Act. *Stanford Law Review* 32(4): 807–826.

Libecap, G.D. 1990. *Contracting for Property Rights.* Cambridge, UK: Cambridge University Press.

Liroff, Richard A. 1980. *Air Pollution Offsets: Trading, Selling and Banking.* Washington, DC: Conservation Foundation.

iski, M., and J.P. Montero. 2005. On Pollution Permit Banking and Market Power. Forthcoming in the *Journal of Regulatory Economics.*

Ludwig, F.L., H.S. Javitz, and A. Valdes. 1983. How Many Stations are Required to Estimate the Design Value and the Expected Number of Exceedances of the Ozone Standard in an Urban Area? *Journal of the Air Pollution Control Association* 33(10): 963–967.

Lyon, R.M. 1982. Auctions and Alternative Procedures for Allocating Pollution Rights. *Land Economics* 58(1): 16.

———. 1986. Equilibrium Properties of Auctions and Alternative Procedures for Allocation Transferable Permits. *Journal of Environmental Economics and Management* 13(2): 129–152.

———. 1989. Transferable Discharge Permit Systems and Environmental Management in Developing Countries. *World Development* 17(8): 1299.

——. 1990. Regulating Bureaucratic Polluters. *Public Finance Quarterly* 2: 198.

Maeda, A. 2003. The Emergence of Market Power in Emission Rights Markets: The Role of Initial Permit Distribution. *Journal of Regulatory Economics* 24(3): 293–314.

Mæstad, O. 2001. Efficient Climate Policy with Internationally Mobile Firms. *Environmental and Resource Economics* 19(3): 267–284.

Maleug, D.A. 1989. Emission Trading and the Incentive to Adopt New Pollution Abatement Technology. *Journal of Environmental Economics and Management* 16(1): 52–57.

——.1990. Welfare Consequences of Emission Credit Trading Programs. *Journal of Environmental Economics and Management* 18(1): 66–77.

Malik, A.S. 1990. Markets for Pollution Control When Firms are Non-Compliant. *Journal of Environmental Economics and Management* 18(2): 97–106.

——. 1992. Enforcement Costs and the Choice of Policy Instruments for Pollution Control. *Economic Inquiry* 30: 714–721.

——. 2002. Further Results on Permit Markets with Market Power and Cheating. *Journal of Environmental Economics and Management* 44(3): 371–390.

Maloney, M.T., and G.L. Brady. 1988. Capital Turnover and Marketable Property Rights. *Journal of Law and Economics* 31(1): 203.

Maloney, M.T., and Robert E. McCormick. 1982. A Positive Theory of Environmental Quality Regulations. *Journal of Law and Economics* 24(1): 99–123.

Maloney, M.T., and B. Yandle. 1984. Estimation of the Cost of Air Pollution Control Regulation. *Journal of Environmental Economics and Management* 11(3): 244–263.

Mani, M., and D. Wheeler. 1998. In Search of Pollution Havens? Dirty Industry in the World Economy, 1960-1995. *Journal of Environment and Development* 7(3): September.

McCann, R.J. 1996. Environmental Commodities Markets: "Messy" versus "Ideal" Worlds. *Contemporary Economic Policy* 14(3): 85.

McGartland, A.M. 1984. Marketable Permit Systems for Air Pollution Control: An Empirical Study. Unpublished Ph.D. dissertation, University of Maryland.

McGartland, A.M., and W.E. Oates. 1985. Marketable Permits for the Prevention of Environmental Deterioration. *Journal of Environmental Economics and Management* 12(3): 207–228.

McLean, Brian. 2003. Ex Post Evaluation of the U.S. Sulphur Allowance Programme. Proceedings of the OECD Workshop on Ex Post Evaluation of Tradable Permits: Methodological and Policy Issues. Paris: OECD.

Melnick, R. Shep. 1983. *Regulation and the Courts: The Case of the Clean Air Act.* Washington, DC: Brookings Institution.

Milliman, S.R., and R. Prince. 1989. Firm Incentives to Promote Technological Change in Pollution Control. *Journal of Environmental Economics and Management* 17(3): 247–265.

Misiolek, W.S., and H.W. Elder. 1989. Exclusionary Manipulation of Markets for Pollution Rights. *Journal of Environmental Economics and Management* 16(2): 156–166.

Montero, Juan-Pablo. 1997. Volunteering for Market-Based Environmental Regulation: The Substitution Provision for the $SO_2$ Emissions Trading Program. Eighth Annual Conference of the European Association of Environmental and Resource Economists, June, Tilburg, Netherlands.

———. 1999. Voluntary Compliance with Market-Based Environmental Policy: Evidence from the U.S. Acid Rain Program. *Journal of Political Economy* 107(5): 998–1033.

———. 2001. Multipollutant Markets. *RAND Journal of Economics* 32(4): 762–774.

———. 2002. Permits, Standards, and Technology Innovation. *Journal of Environmental Economics and Management* 44(1): 23–44.

———. 2002. Prices versus Quantities with Incomplete Enforcement. *Journal of Public Economics* 85(3): 435–454.

———. 2005. Pollution Markets with Imperfectly Observed Emissions. Forthcoming in the *RAND Journal of Economics*.

Montero, J.P. et al. 2002. A Market-Based Environmental Policy Experiment in Chile. *Journal of Law and Economics* 45(1, Part 1): 267–287.

Montgomery, W. David. 1972. Markets in Licenses and Efficient Pollution Control Programs. *Journal of Economic Theory* 5(3): 395–418.

Mrozek, J.R., and A.G. Keeler 2004. Pooling of Uncertainty: Enforcing Tradable Permits Regulation when Emissions are Stochastic. *Environmental and Resource Economics* 29(4): 459–481.

Muller, R.A. et al. 2002. Can Double Auctions Control Monopoly and Monopsony Power in Emissions Trading Markets? *Journal of Environmental Economics and Management* 44(1): 70–92.

National Academy of Public Administration. 1994. *The Environment Goes to Market: The Implementation of Economic Incentives for Pollution Control*. Washington, DC: National Academy of Public Administration.

National Commission on Air Quality. 1981. *To Breathe Clean Air*. Washington, DC: U.S. Government Printing Office.

National Research Council Committee to Review Individual Fishing Quotas. 1999. *Sharing the Fish: Toward a National Policy on Fishing Quotas*. Washington, DC: National Academy Press.

Naughton, M. 1994. Establishing Interstate Markets for Emission Trading of Ozone Precursors. *New York University Environmental Law Journal* 3: 195–249.

Nelson, R. et al. 1993. Differential Environmental Regulation: Effects on Electric Utility Capital Turnover and Emissions. *Review of Economics and Statistics* 75(2): 368.

Nussbaum, B.D. 1992. Phasing Down Lead in Gasoline in the U.S.: Mandates, Incentives, Trading and Banking. In *Climate Change: Designing a Tradable Permit System*, edited by T. Jones and J. Corfee-Morlot. Paris: Organisation for Economic Co-operation and Development, 21–34.

Oates, W.E., and A.M. McGartland. 1985. Marketable Pollution Permits and Acid Rain Externalities: A Comment and Some Further Evidence. *Canadian Journal of Economics* 18(3): 668.

Oehmke, James F. 1987. The Allocation of Pollutant Discharge Permits by Competitive Auction. *Resources and Energy* 9: 153–162.

Office of Air and Radiation. 2004. NO$_x$ Budget Trading Program 2003 Progress and Compliance Report. Report EPA-430-R-04-010. Washinton, DC: U.S. Environmental Protection Agency.

O'Neil, William B. 1980. Pollution Permits and Markets for Water Quality. Unpublished Ph.D. dissertation. University of Wisconsin-Madison.

——. 1983. Transferable Discharge Permit Trading Under Varying Stream Conditions: A Simulation of Multiperiod Permit Market Performances on the Fox River, Wisconsin. *Water Resources Research* 19(3): 608–613.

Organisation for Economic Co-operation and Development. 1992. *Climate Change: Designing a Tradeable Permit System.* Paris: Organisation for Economic Co-operation and Development.

O'Ryan, R. 1996. Cost-Effective Policies to Improve Urban Air Quality in Santiago, Chile. *Journal of Environmental Economics and Management* 31(3): 302-313.

Ostrom, E. et al. (eds.). 2002. *The Drama of the Commons.* Washington, DC: National Academy Press.

Palmer, Adele R., William E. Mooz, Timothy H. Quinn, and Kathleen A. Wolf. 1980. Economic Implications of Regulating Chlorofluorocarbon Emissions from Nonaerosol Applications. Report #R-2524-EPA. Santa Monica, CA: The Rand Corporation.

Palmisano, John. 1983. An Evaluation of Emissions Trading. Paper presented at the 76th Annual Meeting of the Air Pollution Control Association, June 23, Atlanta, GA.

Parry, Ian W.H. 2004. Are Emissions Permits Regressive? *Journal of Environmental Economics and Management* 47: 364–387.

Parry, Ian, and Roberton Williams III. 1999. A Second-Best Evaluation of Eight Policy Instruments to Reduce Carbon Emissions. *Resource and Energy Economics* 21: 347–373.

Parry, Ian, Roberton Williams III, and Lawrence Goulder. 1999. When Can Carbon Abatement Policies Increase Welfare? The Fundamental Role of Distorted Factor Markets. *Journal of Environmental Economics and Management* 37:52–84.

Penderson, Sigurd Lauge. 2003. Experience Gained with $CO_2$ Cap and Trade in Denmark. Proceedings of the OECD Workshop on Ex Post Evaluation of Tradable Permits: Methodological and Policy Issues, January 21–22, Paris.

Pierce, D.F., and P.D. Gutfreund. 1975. Evidentiary Aspects of Air Dispersion Modeling and Air Quality Measurements in Environmental Litigation and Administrative Proceedings. *Federation of Insurance Council Quarterly* 25(Spring): 341–353.

Pigou, A.C. 1920. *The Economics of Welfare.* London: Macmillan.

Plott, Charles R. 1982. Industrial Organization Theory and Experimental Economics. *Journal of Economic Literature* 20(4): 1485–1527.

Popp, D. 2003. Pollution Control Innovations and the Clean Air Act of 1990. *Journal of Policy Analysis and Management* 22(4): 641–660.

Portney, Paul R. 1981. The Macroeconomic Impact of Federal Environmental Regulations. In *Environmental Regulation and the U.S. Economy,* edited by Henry M. Peskin, Paul R. Portney, and Allen V. Kneese. Baltimore, MD: Johns Hopkins University Press for Resources for the Future, 25–54.

Probst, G.L., and R.E. Becker, Jr. 1982. Escaping the Regulatory Dust Bowl: Fugitive Dust and the Clean Air Act. *Natural Resources Lawyer* 14(3): 541–565.

Quinn, Timothy H. 1982. Distributive Consequences and Political Concerns: On the Design of Feasible Market Mechanisms for Environmental Control. In *Buying a Better Environment: Cost-Effective Regulation Through Permit Trading,* edited by Erhard F. Joeres and Martin H. David. Madison, WI: University of Wisconsin Press, 39–54.

Raymond, L. 2003. *Private Rights in Public Resources: Equity and Property Allocation in Market-Based Environmental Policy*. Washington, DC: Resources for the Future.

Reilly, John, Monika Mayer, and Jochen Harnisch. 2000. Multiple Gas Control under the Kyoto Agreement. MIT Joint Program on the Science and Policy of Global Change. Report No. 58.

Rico, R. 1995. The U.S. Allowance Trading System for Sulfur Dioxide: An Update on Market Experience. *Environmental and Resource Economics* 5(2): 115.

Roach, Fred, Charles Kolstad, Allen V. Kneese, Richard Tobin, and Michael Williams. 1981. Alternative Air Quality Policy Options in the Four Corners Region. *Southwestern Review* 1(2): 29–58.

Rose, K.J. 1997. Implementing an Emissions Trading Program in an Economically Regulated Industry: Lessons from the $SO_2$ Trading Program. In *Market-Based Approaches to Environmental Policy: Regulatory Innovations to the Fore*, edited by R. F. Kosobud and J. M. Zimmerman. New York: Van Nostrand Reinhold, 101–136.

Rose, Marshall. 1973. Market Problems in the Distribution of Emission Rights. *Water Resources Research* 9(5): 1132–1144.

Rubin, Jonathan. 1996. A Model of Intertemporal Emission Trading, Banking and Borrowing. *Journal of Environmental Economics and Management* 31(3): 269–286.

Russell, Clifford S. 1981. Controlled Trading of Pollution Permits. *Environmental Science and Technology* 15(1): 1–5.

———. 1990. Game Models for Structuring Monitoring and Enforcement Systems. *Natural Resource Modeling* 4(2): 143–173.

Russell, C.S. et al. 1986. *Enforcing Pollution Control Laws*. Washington, DC: Resources for the Future.

Sand, P.H. 1991. International Cooperation: The Environmental Experience. In *Preserving the Global Environment: The Challenge of Shared Leadership*, edited by J.T. Mathews. New York: W.W. Norton & Co., 236-279.

SanMartin, R. 2003. Marketable Emission Permits with Imperfect Monitoring. *Energy Policy* 31(13): 1369–1378.

Sartzetakis, E.S. 1997a. Raising Rivals' Costs Strategies via Emission Permits Markets. *Review of Industrial Organization* 12(5-6): 751–765.

———. 1997b. Tradeable Emission Permits Regulations in the Presence of Imperfectly Competitive Product Markets: Welfare Implications. *Environmental and Resource Economics* 9(1): 65–81.

Sartzetakis, E.S., and D.G. McFetridge. 1999. Emissions Permits Trading and Market Structure. In *Environmental Regulation and Market Power*, edited by E. Petrakis, E.S. Sartzetakis, and A. Xepapadeas. Cheltenham, UK: Edward Elgar Publishing, 47–66.

Scharer, B. 1999. Tradable Emission Permits in German Clean Air Policy: Considerations on the Efficiency of Environmental Policy Instruments. In *Pollution for Sale: Emissions Trading and Joint Implementation*, edited by S. Sorrell and J. Skea. Cheltenham, UK: Edward Elgar Publishing, 141–153.

Schennach, S.M. 2000. The Economics of Pollution Permit Banking in the Context of Title IV of the 1990 Clean Air Act Amendments. *Journal of Environmental Economics and Management* 40(3): 189–210.

Sedjo, R.A., and G. Marland. 2003. Inter-trading Permanent Emissions Credits and Rented Temporary Carbon Emissions Offsets: Some Issues and Alternatives. *Climate Policy* 3(4): 435–444.

Seskin, Eugene P., Robert J. Anderson, Jr., and Robert O. Reid. 1983. An Empirical Analysis of Economic Strategies for Controlling Air Pollution. *Journal of Environmental Economics and Management* 10(2): 112–124.

Smith, S., and A.J. Yates. 2003. Optimal Pollution Permit Endowments in Markets with Endogenous Emissions. *Journal of Environmental Economics and Management* 46(3): 425–446.

Solomon, B.D., and H.S. Gorman. 1998. State-level Air Emissions Trading: The Michigan and Illinois Models. *Journal of the Air & Waste Management Association* 48(12): 1156–1165.

Sorrell, S. 1999. Why Sulphur Trading Failed in the UK. In *Pollution for Sale. Emissions Trading and Joint Implementation*, edited by S. Sorrell and J. Skea. Cheltenham, UK: Edward Elgar Publishing, 170–210.

South Coast Air Quality Management District. 2001. White Paper on Stabilization of $NO_x$ RTC Prices. Diamond Bar, CA: South Coast Air Quality Management District.

———. 2004. Annual RECLAIM Audit Report for the 2002 Compliance Year. Diamond Bar, CA: South Coast Air Quality Management District.

Spofford, Walter O., Jr., Clifford S. Russell, and Charles M. Paulsen. 1984. *Economic Properties of Alternative Source Control Policies: An Application to the Lower Delaware Valley.* Washington, DC: Resources for the Future.

Stavins, R.N. 1995. Transaction Costs and Tradable Permits. *Journal of Environmental Economics and Management* 29(2): 133–148.

Stavins, R., and R. Hahn. 1993. *Trading in Greenhouse Permits: A Critical Examination of Design and Implementation Issues.* Cambridge, MA: John F. Kennedy School of Government, Harvard University.

Stevens, B. and A. Rose. 2002. A Dynamic Analysis of the Marketable Permits Approach to Global Warming Policy: A Comparison of Spatial and Temporal Flexibility. *Journal of Environmental Economics and Management* 44(1): 45–69.

Stranlund, J.K., and C.A. Chavez. 2000. Effective Enforcement of a Transferable Emissions Permit System with a Self-Reporting Requirement. *Journal of Regulatory Economics* 18(2): 113–131.

Stranlund, J.K., and K.K. Dhanda. 1999. Endogenous Monitoring and Enforcement of a Transferable Emissions Permit System. *Journal of Environmental Economics and Management* 38(3): 267–282.

Stranlund, J.K., C.A. Chavez, and B.C. Field. 2002. Enforcing Emissions Trading Programs: Theory, Practice, And Performance. Paper presented at the 2nd CATEP Workshop on the Design and Integration of National Tradable Permit Schemes for Environmental Protection, March 25–26, University College, London.

Svendsen, G.T. 1999. Interest Groups Prefer Emission Trading: A New Perspective. *Public Choice* 101(1-2): 109–128.

Taylor, M.R., E.S. Rubin, and D. Hounshell. 2005. Regulation as the Mother of Invention: The Case of $SO_2$ Control. *Law & Policy* 27(2): 348–378.

Teller, Azriel. 1970. Air Pollution Abatement: Economic Rationality and Reality. In *America's Changing Environment*, edited by Roger Revelle and Hans Landsberg. Boston: Beacon Press, 39–55.

Tietenberg, Thomas H. 1973. Controlling Pollution by Price and Standards Systems. *Swedish Journal of Economics* 75(2): 193–203.

Tietenberg, T. H. 1974. The Design of Property Rights for Air Pollution Control. *Public Policy* 27(3): 275.

———. 1985. *Emissions Trading: An Exercise in Reforming Pollution Policy*. Washington, DC: Resources for the Future.

———. 1990. Economic Instruments for Environmental Regulation. *Oxford Review of Economic Policy* 6(1): 17–33.

———. 1995. Tradable Permits for Pollution Control When Emission Location Matters: What Have We Learned? *Environmental and Resource Economics* 5(2): 95–113.

———. 1998. Ethical Influences on the Evolution of the US Tradeable Permit Approach to Pollution Control. *Ecological Economics* 24(2,3): 241–257.

———. 2002. The Tradable Permits Approach to Protecting the Commons: What Have We Learned? In *The Drama of the Commons*, edited by N.R. Council. Washington, DC: National Academy Press, 197–232.

Tietenberg, T. et al. 1998. International Rules for Greenhouse Gas Emissions Trading: Defining the Principles, Modalities, Rules and Guidelines for Verification, Reporting and Accountability. UNCTAD/GDS/GFSB/Misc.6. Geneva: United Nations.

Tirole, J. 1988. *The Theory of Industrial Organization*. Cambridge, MA: MIT Press.

Toman, M., and K. Palmer. 1997. How Should an Accumulative Substance be Banned? *Environmental and Resource Economics* 9(1): 83–102.

U.S. Environmental Protection Agency (U.S. EPA). 1982. Third Quarter Report of the Economic Dislocation Early Warning System. Washington, DC: U.S. EPA.

———. 1997. The Benefits and Costs of the Clean Air Act, 1970–1990. Washington, DC: U.S. EPA.

———. 2002. An Evaluation of the South Coast Air Quality Management District's Regional Clean Air Incentives Market—Lessons in Environmental Markets and Innovation. San Francisco, CA: U.S. EPA Region 9.

———. 2004. $NO_x$ Budget Trading Program: 2003 Progress and Compliance Report. Report EPA-430-R-04-010. Washington, DC: U.S. EPA.

U.S. General Accounting Office (U.S. GAO). 1979. Improvements Needed in Controlling Major Air Pollution Sources. Report CED-78-165. Washington, DC: U.S. GAO.

———. 1981. Clean Air Act: Summary of GAO Reports, October 1977 through January 1981, and Ongoing Reviews. Report CED-81-84. Washington, DC: U.S. GAO.

———. 1982. Cleaning Up the Environment: Progress Achieved but Major Unresolved Issues Remain. Report CED-82-72. Washington, DC: U.S. GAO.

———. 2001. Air Pollution: EPA Should Improve Oversight of Emissions Reporting by Large Facilities. Report GAO-01-46. Washington, DC: U.S. GAO.

Van Egteren, H., and M. Weber. 1996. Marketable Permits, Market Power and Cheating. *Journal of Environmental Economics and Management* 30(2): 161–173.

Vickrey, William. 1961. Counterspeculation, Auctions, and Competitive Sealed Tenders. *Journal of Finance* 16(March): 8–37.

Vivian, W. 1981. An Updated Tabulation of Offset Cases. Cited in National Commission on Air Quality, *To Breathe Clean Air*. Washington, DC: U.S. Government Printing Office.

Vivian, W., and W. Hall. 1979. *An Empirical Examination of U.S. Market Trading in Air Pollution Offsets*. Ann Arbor, MI: Institute of Public Policy Studies, University of Michigan.

von der Fehr, N. 1993. Tradable Emission Rights and Strategic Interaction. *Environmental and Resource Economics* 3: 129–151.

Wasserman, Cheryl et al. 1992. *Principles of Environmental Enforcement*. Washington, DC: U.S. EPA.

Weitzman, M.L. 1974. Prices vs. Quantities. *Review of Economic Studies* 41: 477–491.

Westkog, H. 1996. Market Power in a System of Tradable $CO_2$ Quotas. *The Energy Journal* 17: 85–103.

Wiener, J. 2004. Hormesis, Hotspots and Emissions Trading. *Human and Experimental Toxicology* 23(6): 289–301.

Williams, R.C. III. 2002. Prices vs. Quantities vs. Tradable Quantities. Working Paper 9283. Cambridge, MA: National Bureau of Economic Research.

———. 2003. Cost-Effectiveness vs. Hot Spots: Determining the Optimal Size of Emissions Permit Trading Zones. University of Texas at Austin Working Paper.

Working Group III. 2001. *Climate Change 2001: Mitigation*. Cambridge, UK: Intergovernmental Panel on Climate Change by the Cambridge University Press.

Yaron, Dan. 1979. A Model for the Analysis of Seasonal Aspects of Water Quality Control. *Journal of Environmental Economics and Management* 6(2): 140–151.

Yates, A.J., and M.B. Cronshaw. 2001. Pollution Permit Markets with Intertemporal Trading and Asymmetric Information. *Journal of Environmental Economics and Management* 42(1): 104–118.

Zhang, Z.X. 1999. International Greenhouse Emissions Trading: Who Should Be Held Liable for the Non-compliance by Sellers? *Ecological Economics* 31(3): 323–329.

Zylicz, T. 1999. Obstacles to Implementing Tradable Pollution Permits: the Case of Poland. In *Implementing Domestic Tradable Permits for Environmental Protection*. Paris: Organisation for Economic Co-operation and Development, 147–165.

# Index

# About the Author

T. H. (Tom) TIETENBERG is the Mitchell Family Professor of Economics at Colby College. He is the author or editor of 11 books and over 100 articles and essays on environmental and natural resource economics, including *Environmental and Natural Resources Economics*, one of the best selling textbooks in the field. A former president of the Association of Environmental and Resource Economists (1987-1988), he was the team leader for the United Nations Project to define how Article 17 of the Kyoto Protocol to the Climate Change Convention, which authorizes emissions trading among the industrialized nations, would be implemented. Tietenberg has consulted with the World Bank, U.S. AID, the Organization for Economic Co-operation and Development, the United Nations and several foreign governments on the use of economics incentives to produce sustainable development. In 2005 he was named a fellow by the Association of Environmental and Resource Economists.